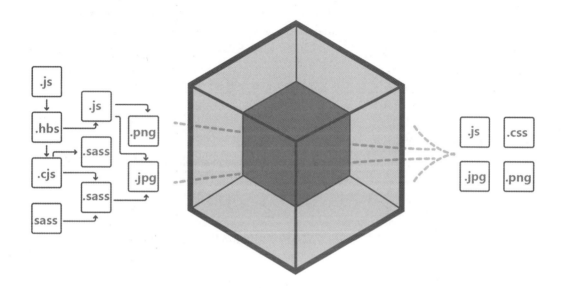

Node.js+Webpack
开发实战

夏磊 著

清华大学出版社
北京

内 容 简 介

Node.js 是一个基于 Chrome V8 引擎的 JavaScript 运行环境，它用于构建高速、可伸缩的网络应用程序，为前端开发提供了新的机遇。为了让前端开发者更有效地使用 Node.js 进行开发，作者结合自己的开发经验编著了本书，全书提供了丰富的示例代码，详细讲述和演示了如何将所学的知识应用于实际的开发中。

本书分为三部分共 21 章，第一部分 Node.js 基础：Node.js 概述，搭建 Node.js 开发环境，Node.js 编程基础；第二部分后端的 Node.js：Express 框架，Koa 框架，MongoDB 数据库，MySQL 数据库，ORM 框架 Sequelize，微博系统实战项目，高性能内存型数据库 Redis，前端的发展现状；第三部分前端的 Node.js：前端发展状况，Webpack 基础，Webpack 常用配置，Webpack 构建 Vue 应用，Webpack 构建 React 应用，服务端渲染技术和同构应用的开发，Webpack 构建传统多页面 Web 应用，Webpack 性能优化，Webpack 自定义 Loader 的编写，Webpack 自定义 Plugin 的编写。

本书适合 Node.js+Webpack 前端开发工程师作为自学参考书，也适合高等院校和培训学校相关专业的师生作为教学参考书。

本书封面贴有清华大学出版社防伪标签，无标签者不得销售。
版权所有，侵权必究。举报：010-62782989，beiqinquan@tup.tsinghua.edu.cn。

图书在版编目（CIP）数据

Node.js+Webpack 开发实战 / 夏磊著. —北京：清华大学出版社，2020.6（2021.8重印）
ISBN 978-7-302-55595-7

Ⅰ. ①N… Ⅱ. ①夏… Ⅲ. ①JAVA 语言—程序设计②网页制作工具—程序设计 Ⅳ. ①TP312.8 ②TP392.092.2

中国版本图书馆 CIP 数据核字（2020）第 089921 号

责任编辑：夏毓彦
封面设计：王　翔
责任校对：闫秀华
责任印制：宋　林

出版发行：清华大学出版社
　　　　网　　　址：http://www.tup.com.cn，http://www.wqbook.com
　　　　地　　　址：北京清华大学学研大厦 A 座　　邮　编：100084
　　　　社　总　机：010-62770175　　邮　购：010-62786544
　　　　投稿与读者服务：010-62776969，c-service@tup.tsinghua.edu.cn
　　　　质　量　反　馈：010-62772015，zhiliang@tup.tsinghua.edu.cn
印 装 者：三河市君旺印务有限公司
经　　销：全国新华书店
开　　本：190mm×260mm　　印　张：20.25　　字　数：518 千字
版　　次：2020 年 8 月第 1 版　　印　次：2021 年 8 月第 2 次印刷
定　　价：69.00 元

产品编号：076384-01

前　　言

Node.js 是一个基于 Chrome V8 引擎的 JavaScript 运行环境，用于构建高速、可伸缩的网络应用程序。

事实上，Node.js 不仅仅用来构建网络应用程序，还为前端开发提供了新的机遇。现在拥有 JavaScript 经验的开发人员可以在前端和后端使用 Node.js，降低了语言导致的过渡成本。Node.js 拥有一个巨大的 JavaScript 生态系统，再加上这几年前端的发展，出现了许多新框架和新语言，但是对于初次接触 Node.js 的用户来说是不太友好的，再加上缺乏系统性的指南，导致前端开发者无法有效地学好 Node.js，作者编著本书希望对改变这种情况尽绵薄之力。

关于本书

第一部分　Node.js 基础

第一部分是对 Node.js 的介绍，涵盖了它的原理和基础知识。

第 1 章介绍 Node.js 的原理和应用场景。

第 2 章介绍如何在计算机上安装 Node.js 以及 Visual Studio Code 编辑器。我们将用一个简单的 HTTP 服务器来测试 Node.js 是否成功安装。

第 3 章介绍 Node.js 的编程基础。内容包括 NPM、模块系统、异步编程方式和常用模块。

第二部分　后端的 Node.js

第二部分是对后端 Node.js 的介绍，涵盖了主流的 Web 框架和常用组件，包含数据库、缓存，等等。

第 4 章详细介绍 Express 开发框架，这是最早也是最流行的 Node.js Web 开发框架。内容包括 Express 的请求路由、请求与响应、中间件、错误处理和页面渲染。最后演示如何使用 Express 框架开发留言板系统。

第 5 章详细介绍 Koa 框架。Koa 框架被称为"下一代的 Web 开发框架"，Koa 的"一切皆为中间件"思想被其他 Web 框架广泛地采用。本章内容包括 Koa 的上下文对象、中间件模型、请求路由、错误处理和模板渲染。最后演示如何使用 Koa 开发博客系统。

第 6 章介绍 MongoDB 数据库。MongoDB 通常被称为 Node.js 的"黄金搭档"，因为 MongoDB 采用了"BeJSON"的结构，对 JavaScript 有天然的亲和性。本章内容包括 MongoDB 的安装、基础使用和 Node.js 对 MongoDB 的操作。

第 7 章介绍 MySQL 数据库，这是目前最流行的、开源的关系型数据库系统。内容包括

MySQL 的安装、基础语法、关联关系和事务操作，为后续的实战项目打下基础。

第 8 章介绍 ORM 框架——Sequelize，Sequelize 一个操作 MySQL 的框架，能够通过对象的方式操作数据库。本章内容包括 Sequelize 模型、关联关系、对数据的操作和事务的使用。

第 9 章介绍一个完整的实战项目开发过程。我们将基于 Koa 和 Sequelize 来开发一个微博系统，带领大家学习一个完整项目的研发流程。

第 10 章介绍高性能内存型 NoSQL 数据库 Redis，Redis 常用在高并发场景，比如秒杀活动、抽奖、排行榜等。本章内容包括 Redis 的基础知识，数据结构以及 Node.js 对 Redis 的操作。

第 11 章介绍实时 Web 通信技术 WebSocket，WebSocket 的出现赋予了 Web 应用更多的可能性。本章内容包括传统的实时 Web 技术、WebSocket 协议的原理以及使用 Node.js 实现 WebSocket 服务器，最后演示如何使用 Node.js 来构建一个在线聊天室。

第三部分　前端的 Node.js

第三部分介绍前端的发展以及 Node.js 在前端的应用，重点介绍目前最流行的构建工具——Webpack。

第 12 章介绍前端的发展现状，包括模块系统、新语言、新框架和新的构建工具。

第 13 章介绍 Webpack 的基础使用和核心概念，包括如何一步一步对 Webpack 进行配置以及 Loader 和 Plugin 的使用。

第 14 章详细介绍 Webpack 的常用配置。

第 15 章介绍如何使用 Webpack 构建 Vue 应用，包括构建 Vue 应用需要的模块、相应的配置和导入 TypeScript 支持。

第 16 章介绍如何使用 Webpack 构建 React 应用，包括 JSX 语法、Babel 工具、Webpack 的配置以及导入 TypeScript 支持。

第 17 章介绍服务端渲染技术和同构应用的开发，包括服务端渲染技术的原理以及如何使用 Webpack 构建同构应用，最后演示如何构建一个 React 的同构应用。

第 18 章详细介绍如何使用 Webpack 构建传统多页面 Web 应用。

第 19 章详细介绍 Webpack 性能优化的常用手段，包括优化配置、提取公共代码、多进程编译、按需加载和热更新的知识。

第 20 章介绍 Webpack 自定义 Loader 的编写，包括基本 Loader、Loader 配置、异步 Loader 等知识，最后演示如何编写一个多语言 Loader。

第 21 章介绍 Webpack 自定义 Plugin 的编写，包括 Webpack 构建流程、Compiler 和 Compilation、Tapable 对象和常用 API，最后演示清单文件插件的编写以及将构建结果上传到 CDN 插件的编写。

示例代码下载

本书提供了丰富的示例，演示如何利用每个所学的知识点。本书的源码已经托管到 GitHub 网站，读者可以通过 https://github.com/nodejs-inaction 链接进行访问及下载。也可以扫描下面的二维码下载。

如果你在下载过程中遇到问题，可发送邮件至 booksaga@163.com 获得帮助，邮件标题为"Node.js+Webpack 开发实战"。

关于作者

夏磊，毕业于湖南工业大学网络工程专业，拥有多年研发经验。在过去的几年里，他还是很多开源项目的贡献者。精通 PHP 脚本开发、Node.js/Golang 服务端开发以及 JavaScript 开发，善于把握与应用新技术，博客和公众号上有大量 Web 相关技术文章，深受读者好评。著有图书《ThinkPHP 实战》和《ThinkPHP5 实战》。

著者
2020 年 5 月

目 录

第一部分　Node.js 基础篇

第 1 章　Node.js 概述 3
1.1　Node.js 是什么 3
1.2　Node.js 的运行原理 5
1.3　Node.js 的应用场景 6
　　1.3.1　Node.js 优缺点 6
　　1.3.2　应用场景 7
1.4　本章小结 7

第 2 章　搭建 Node.js 开发环境 8
2.1　安装 Node.js 8
　　2.1.1　Windows 上安装 Node.js 8
　　2.1.2　Linux 安装 Node.js 9
　　2.1.3　Ubuntu 安装 Node.js 9
　　2.1.4　CentOS 安装 Node.js 9
　　2.1.5　macOS 安装 Node.js 10
2.2　安装 VSCode 编辑器 10
2.3　编写 HTTP 服务器 11
2.4　本章小结 12

第 3 章　Node.js 编程基础 13
3.1　NPM 包管理器介绍 13
　　3.1.1　更换 NPM 镜像源 13
　　3.1.2　初始化项目 14
　　3.1.3　使用 npm 命令安装模块 14
　　3.1.4　本地安装与全局安装 14
　　3.1.5　生产依赖和开发依赖 15
　　3.1.6　其他 npm 命令 15
3.2　Yarn 包管理器介绍 16
　　3.2.1　安装 Yarn 16

3.2.2　Yarn 常用命令 16
3.3　解读 package.json 文件 16
　　3.3.1　package.json 字段说明 17
　　3.3.2　版本号说明 18
　　3.3.3　常见的版本号限定符 18
3.4　Node.js 的模块系统 18
　　3.4.1　module 和 exports 19
　　3.4.2　require 20
　　3.4.3　开发一个自定义模块 21
3.5　Node.js 的异步编程风格 22
　　3.5.1　回调函数 22
　　3.5.2　Promise 23
　　3.5.3　async/await 26
3.6　Node.js 常用核心模块 28
　　3.6.1　events 模块 28
　　3.6.2　fs 模块 29
　　3.6.3　stream 接口 32
　　3.6.4　http 模块 35
3.7　本章小结 36

第二部分　后端的 Node.js

第 4 章　最流行的 Web 框架——Express 39
4.1　框架简介 39
4.2　快速开始 40
　　4.2.1　初始化项目 40
　　4.2.2　开始编码 40
　　4.2.3　运行应用 41
　　4.2.4　小结 41
4.3　路由 41
　　4.3.1　路由方法 42
　　4.3.2　路由路径 42
　　4.3.3　路由参数 44
　　4.3.4　路由函数 45
4.4　请求对象 47
　　4.4.1　获取请求 Cookie 49
　　4.4.2　获取请求体 50
4.5　响应对象 50
4.6　中间件 54

	4.6.1	全局中间件 54
	4.6.2	路由中间件 55
	4.6.3	可配置的中间件 55
	4.6.4	Cookie 中间件 57
	4.6.5	响应时长中间件 57
	4.6.6	静态资源中间件 58
4.7	错误处理 59
	4.7.1	同步错误 59
	4.7.2	异步错误 60
	4.7.3	自定义错误处理函数 60
	4.7.4	多个错误处理函数 61
4.8	模板渲染 62
	4.8.1	使用 ejs 模板 62
	4.8.2	ejs 语法 63
4.9	留言板项目开发 65
	4.9.1	开始编码 65
	4.9.2	运行项目 67
4.10	本章小结 68

第 5 章 下一代 Web 开发框架——Koa 70

5.1	Koa 简介 70
5.2	Bluebird 71
5.3	Koa 快速开始 72
	5.3.1 初始化项目 72
	5.3.2 开始编码 73
5.4	Context 73
5.5	Cookie 操作 75
	5.5.1 Cookie 签名 75
	5.5.2 写入 Cookie 75
	5.5.3 读取 Cookie 76
	5.5.4 中间件 76
	5.5.5 请求日志中间件 78
	5.5.6 可配置的中间件 79
	5.5.7 Cookie 解析中间件 80
	5.5.8 路由函数 81
	5.5.9 多个路由函数 81
	5.5.10 错误处理 82
	5.5.11 多个错误处理器 83
5.6	路由系统 84

5.6.1	快速开始	85
5.6.2	路由对象	85
5.6.3	路由路径	86
5.6.4	路由函数	86
5.6.5	路由级别中间件	87
5.6.6	路由前缀	87
5.6.7	模块化路由	88

5.7 模板渲染 89
  5.7.1 快速开始 89
  5.7.2 模板布局 90

5.8 博客项目实战 92
  5.8.1 功能梳理 92
  5.8.2 项目代码 93
  5.8.3 效果展示 100
  5.8.4 项目小结 102

5.9 本章小结 102

第 6 章 文档型 NoSQL 数据库——MongoDB 103

6.1 简介 103
  6.1.1 主要特点 103
  6.1.2 概念 104
  6.1.3 数据库 104
  6.1.4 集合 105
  6.1.5 文档 105

6.2 安装 106
  6.2.1 Windows 106
  6.2.2 Linux 107
  6.2.3 macOS 108

6.3 常用操作 109
  6.3.1 创建数据库 109
  6.3.2 删除数据库 109
  6.3.3 创建集合 110
  6.3.4 查看集合 110
  6.3.5 删除集合 110
  6.3.6 索引 111
  6.3.7 插入文档 112
  6.3.8 更新文档 112
  6.3.9 删除文档 113
  6.3.10 查询文档 113

		6.3.11 其他查询语法	115
6.4	Node.js 集成		116
		6.4.1 初始化项目	116
		6.4.2 连接数据库	116
		6.4.3 mongoose 的关键概念	117
		6.4.4 Schema	117
		6.4.5 Model	120
6.5	本章小结		121

第 7 章 最流行的关系型数据库——MySQL 123

7.1	简介		123
7.2	安装		123
		7.2.1 Windows	124
		7.2.2 Linux	124
		7.2.3 macOS	126
7.3	术语		126
7.4	索引		127
		7.4.1 普通索引	127
		7.4.2 唯一索引	127
		7.4.3 联合索引	128
7.5	事务		128
		7.5.1 ACID 原则	128
		7.5.2 事务并发问题	129
		7.5.3 隔离级别	129
		7.5.4 事务控制语句	129
7.6	关联关系		130
		7.6.1 一对多关联	130
		7.6.2 一对一关联	131
		7.6.3 多对多关联	131
7.7	数据库操作		132
7.8	数据类型		133
7.9	数据表操作		135
		7.9.1 创建数据表	135
		7.9.2 删除数据表	136
		7.9.3 添加字段	136
		7.9.4 删除字段	137
		7.9.5 修改字段	137
7.10	数据操作		137
		7.10.1 插入数据	137

	7.10.2 查询数据	138
	7.10.3 修改数据	139
	7.10.4 删除数据	139
7.11	本章小结	140

第 8 章 ORM 框架——Sequelize 141

8.1	ORM	141
8.2	Sequelize 简介	142
8.3	快速开始	142
8.4	构造方法	143
8.5	数据类型	144
8.6	模型定义	146
	8.6.1 字段设置	146
	8.6.2 模型选项	147
	8.6.3 Hooks	148
	8.6.4 生命周期函数	149
	8.6.5 模型验证器	150
	8.6.6 模型方法	155
	8.6.7 索引	156
	8.6.8 数据库同步	157
8.7	模型使用	157
	8.7.1 插入数据	158
	8.7.2 更新数据	159
	8.7.3 删除数据	160
	8.7.4 查询数据	161
	8.7.5 查询语法	164
	8.7.6 事务	165
8.8	关联	167
	8.8.1 hasOne	167
	8.8.2 belongsTo	169
	8.8.3 hasMany	171
	8.8.4 belongsToMany	173
8.9	本章小结	175

第 9 章 微博项目开发 176

9.1	功能分析	176
9.2	数据模型	177
9.3	开始编码	177
	9.3.1 初始化项目	177

		9.3.2	项目目录	178
		9.3.3	路由设计	178
		9.3.4	共享组件	178
		9.3.5	中间件	179
		9.3.6	模型代码	180
		9.3.7	生成数据表	183
		9.3.8	业务代码	184
		9.3.9	路由代码	188
		9.3.10	视图文件	192
		9.3.11	Web 应用引导文件	193
	9.4	效果展示		194
	9.5	项目代码		196
	9.6	本章小结		196

第 10 章 高性能内存型 NoSQL 数据库——Redis ... 197

	10.1	Redis 简介		197
		10.1.1	特点	197
		10.1.2	应用场景	198
	10.2	Redis 安装		198
		10.2.1	在 Windows 下安装 Redis	198
		10.2.2	在 Linux 下安装 Redis	199
		10.2.3	在 macOS 下安装 Redis	199
	10.3	Redis 支持的数据结构		200
		10.3.1	String（字符串）	200
		10.3.2	哈希表（Hash）	201
		10.3.3	列表（List）	202
		10.3.4	集合（Set）	203
		10.3.5	有序集合（ZSet）	203
		10.3.6	发布订阅	204
	10.4	Node.js 集成 Redis		205
		10.4.1	快速开始	205
		10.4.2	Promise	206
	10.5	本章小结		207

第 11 章 实时双向 Web 技术——WebSocket ... 208

	11.1	传统的实时 Web 技术		208
		11.1.1	Ajax 轮询（Ajax Polling）	208
		11.1.2	服务器推送（Comet）	209
	11.2	WebSocket		209

11.3 实现 WebSocket 握手协议 210
    11.3.1 握手协议过程 211
    11.3.2 服务端代码 211
    11.3.3 客户端代码 212
11.4 使用 ws 模块开发聊天室 212
    11.4.1 安装依赖 213
    11.4.2 服务端代码 213
    11.4.3 客户端代码 214
11.5 本章小结 215

第三部分　前端中的 Node.js

第 12 章　迅速发展的前端技术 219
12.1 模块系统 219
    12.1.1 CommonJS 220
    12.1.2 AMD 220
    12.1.3 CMD 221
    12.1.4 ES6 模块化 221
12.2 新语言 222
    12.2.1 ES6 222
    12.2.2 TypeScript 222
    12.2.3 Less 223
    12.2.4 SCSS 223
12.3 新框架 224
    12.3.1 AngularJS 224
    12.3.2 React 224
    12.3.3 Vue 224
    12.3.4 Angular 225
12.4 构建工具 225
    12.4.1 Grunt 226
    12.4.2 Gulp 226
    12.4.3 Webpack 227
12.5 本章小结 228

第 13 章　Webpack 起步 229
13.1 安装 229
13.2 示例项目 230
13.3 Loader 231
    13.3.1 CSS 处理 231

	13.3.2 图片处理	232
13.4	Plugin	234
	13.4.1 提取 CSS	234
	13.4.2 自动更新 HTML 中的资源引用	236
13.5	开发服务器	237
13.6	核心概念	239
13.7	本章小结	239

第 14 章 Webpack 配置 240

14.1	Mode	241
14.2	Entry 和 Context	241
	14.2.1 不配置 Context 的情况	241
	14.2.2 配置 Context 的情况	242
14.3	Output	242
	14.3.1 chunkFilename	243
	14.3.2 path	243
	14.3.3 publicPath	244
	14.3.4 libraryTarget 和 library	244
14.4	Module	246
	14.4.1 noParse	246
	14.4.2 rules	247
14.5	Resolve	248
	14.5.1 alias	248
	14.5.2 extensions	249
	14.5.3 mainFields	249
	14.5.4 modules	250
14.6	devtool	250
14.7	externals	250
14.8	DevServer	251
14.9	Plugins	252
14.10	完整示例	252
14.11	本章小结	254

第 15 章 Vue 实战 255

15.1	Hello World	255
15.2	配置 Webpack	257
	15.2.1 Loader 和 Plugin	257
	15.2.2 安装依赖模块	257
	15.2.3 编写配置文件	258

	15.2.4 执行构建	259
15.3	生产构建	259
	15.3.1 Webpack 配置	259
	15.3.2 package.json 修改	260
15.4	TypeScript 支持	261
	15.4.1 TypeScript 配置	261
	15.4.2 Webpack 配置	262
	15.4.3 App.vue	263
15.5	本章小结	264

第 16 章 React 实战 265

16.1	JSX	265
16.2	Babel	266
16.3	TypeScript	268
16.4	本章小结	271

第 17 章 服务端渲染 272

17.1	SSR 原理	273
17.2	添加 SSR 的 webpack.config.js	273
17.3	添加 SSR 的入口文件	274
17.4	添加 SSR 打包命令	275
17.5	执行构建	275
17.6	添加 Node.js HTTP 服务器	275
17.7	目录结构	276
17.8	运行应用	276
17.9	本章小结	277

第 18 章 多页应用脚手架 278

18.1	项目结构	278
18.2	开发步骤	279
	18.2.1 初始化项目与安装依赖	279
	18.2.2 配置	280
18.3	业务代码	282
18.4	本章小结	283

第 19 章 性能优化 284

19.1	限定 Webpack 处理文件范围	284
19.2	DllPlugin	285
19.3	HappyPack	287

19.4	Tree-Shaking	288
19.5	按需加载	289
19.6	提取公共代码	289
19.7	热更新	290
19.8	本章小结	290

第 20 章 编写自定义 Loader ... 291

20.1	基本 Loader	291
20.2	Loader 选项	293
20.3	异步 Loader	294
20.4	"Raw" Loader	295
20.5	读取 Loader 配置文件	295
	20.5.1 项目结构	295
	20.5.2 执行构建	297
20.6	本章小结	297

第 21 章 编写自定义插件 ... 298

21.1	基本构建流程	298
21.2	插件示例	299
21.3	Compiler 与 Compilation 对象	299
21.4	Tapable	300
21.5	常用操作	301
	21.5.1 读取输出资源、模块及依赖	301
	21.5.2 修改输出资源	302
21.6	插件编写实例	302
	21.6.1 生成清单文件	303
	21.6.2 构建结果上传到 CDN	304
21.7	本章小结	306

第一部分 Node.js基础篇

Node.js 是一个基于 Chrome V8 引擎的 JavaScript 运行环境，用于构建高速、可伸缩的网络应用程序。

在这一部分，我们将学习 Node.js 的原理、应用场景、开发环境的搭建和 Node.js 的基础知识。

第 1 章

Node.js 概述

本章内容

- Node.js 是什么
- Node.js 的模块架构
- Node.js 的运行原理
- Node.js 的应用场景

1.1 Node.js 是什么

你很有可能已经听说过 Node.js，甚至已经用上了。的确，近年来不管是 Web/服务端还是桌面/移动应用的开发都可以看到它的身影。它在 GitHub 上拥有 64000 多颗星（Star），拥有 2500 多位贡献者，官方的包管理平台 NPM 下拥有高达 100 多万的软件包（Package）。所有这些都表明 Node.js 的强大活力。

官网上（https://nodejs.org）给 Node.js 的定义是：

Node.js 是一个基于 Chrome V8 引擎的 JavaScript 运行环境，用于构建高速、可伸缩的网络应用程序。Node.js 采用的事件驱动、非阻塞 I/O 模型，使它既轻量又高效，是构建可扩展的网络应用程序的完美选择。

1. JavaScript 引擎

JavaScript 引擎是一个专门处理 JavaScript 脚本的虚拟机，一般会附带在网页浏览器之中。目前为止一些知名的 JavaScript 引擎如下：

- V8 引擎：Chrome 浏览器采用的 JavaScript 引擎，也是第一个使用 JIT 技术的引擎。

- JavaScriptCore：Safari 浏览器采用的 JavaScript 引擎。
- Chakra：IE/Edge 浏览器采用的 JavaScript 引擎。
- SpiderMonkey：Mozilla Firefox 浏览器采用的 JavaScript 引擎。

为什么会有这么多的 JavaScript 引擎呢？

主要是因为历史的原因，1995 年，网景（NetScape）公司开发了一种运行在浏览器的脚本语言，最初命名为 Mocha，后来改名为 LiveScript，临近发布的时候为了蹭一蹭 Java 的热度，最终命名为 JavaScript。直到 1997 年，JavaScript 才被欧洲计算机制造商协会（ECMA）进行了标准化，标准编号为 ECMA-262，标准化后的名称为 ECMAScript。

ECMAScript 只是一种语言规范，满足该规范的语言都可以称为"ECMAScript 兼容"。现在比较出名的是 JavaScript 语言，早些年还有 ActionScript（Flash 使用的脚本语言）。

2. V8 引擎

V8 是一个由 Google 公司开发的开源 JavaScript 引擎，使用 C++编写，用于 Google Chrome 浏览器和 Google Chromium 浏览器。

V8 在运行之前将 JavaScript 编译成了机器代码，而非字节码或是解释执行它，以此提升性能。此外，V8 还使用了如内联缓存（Inline Caching）等技术来提高性能。有了这些功能，JavaScript 程序在 V8 引擎上的速度可以媲美二进制编译程序的执行速度。

Node.js 也因为采用了 V8 引擎才有如此高的 JavaScript 执行效率。

3. 事件驱动

事件驱动程序设计（Event-Driven Programming）是一种计算机程序设计模型。这种模型的程序运行流程是由用户的操作（如鼠标的按键、键盘的按键操作）或者是由其他程序的消息来驱动的。

事件驱动的程序至少会有一个事件队列，当有新的请求进来时会被插入到队列中，然后通过循环来检测队列中的事件，当发现有一个事件发生时就会调用回调函数。

Node.js 的 JavaScript（JS）线程是单线程运行的，通过一个事件循环（Event Loop）来循环取出消息队列（Event Queue）中的消息进行处理，处理过程基本上就是去调用该消息对应的回调函数。

4. 非阻塞 I/O

I/O 模型的一种，与之相对的还有阻塞 I/O。

简单来说，阻塞 I/O 就是进行 I/O 操作的时候，进程会被阻塞，直到 I/O 操作完成后进程才会解除阻塞状态继续执行。而非阻塞 I/O 在进行 I/O 操作时，进程不会被阻塞，进程继续执行；如果需要知道 I/O 操作的结果，可以通过轮询的方式。

5. Node.js 的模块架构

Node.js 核心基于 libuv 框架，由 C/C++编写。图 1-1 所示是 Node.js 的模块架构图。

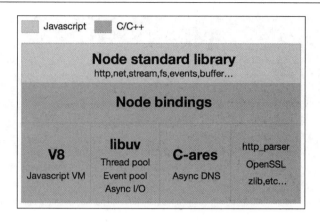

图 1-1

6. Node standard library（Node 标准库）

Node.js 标准库由 JavaScript 编写而成，它提供网络、文件、事件等操作，可以认为是一层比较薄的 API 封装层，实际的操作还是由底层来完成。

7. Node bindings（Node 绑定层）

Node.js 底层使用了大量的 C/C++代码，是高性能的保证。不过，JavaScript 代码是怎么和这些 C/C++代码相互调用的呢？这里不是使用了好几种语言吗？

一般来说，不同语言之间的规范不同，所以写出来的代码无法直接沟通，这时候就需要 Bindings 层，它是一些胶水代码，能够把不同的语言绑定在一起使其能够相互沟通。在 Node.js 中，Bindings 层所做的就是把底层的 C/C++接口暴露给 JavaScript 环境，从而打通 JavaScript 与 C/C++之间的相互调用。

8. libuv

libuv 是提供异步功能的 C 库。它在运行时负责一个事件循环（Event Loop）、一个线程池、文件系统 I/O、DNS 相关的 I/O 和网络 I/O，以及一些其他重要功能。

9. 其他的 C/C++组件和库

如 C-ares、http_parser、OpenSSL 以及 zlib 等，这些依赖提供了对系统底层功能的访问，比如网络、压缩、加解密，等等。

1.2 Node.js 的运行原理

我们已经了解了 Node.js 组成部分各自的面貌。现在来看看它们是如何相互协作以提供强大的网络服务功能。Node.js 系统如图 1-2 所示。

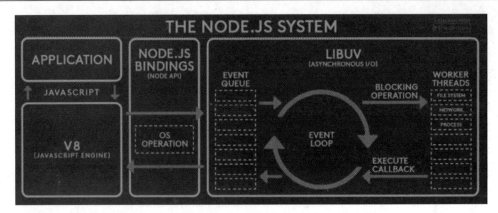

图 1-2

Node.js 应用启动时，会开启 JS 线程（主线程）、由 libuv 提供的线程池（Worker Threads）和一个事件循环（Event Loop）。JS 线程负责执行应用代码，当发现有 I/O 操作时，直接提交给 libuv 的线程池并注册回调函数，不会等待 I/O 结束后再继续运行，而是拿到一个状态后继续执行，这就是"单线程非阻塞 I/O"。

I/O 操作结束之后会有一个事件，该事件会放在事件队列（Event Queue）中，事件循环每次都会检查是否有事件需要处理，如果有就处理，否则进行下一轮轮询；如果没有任何事件需要处理则退出进程。这就是"事件驱动"。

在 I/O 密集型应用中，主线程只负责提交任务，轮询结果，耗时的任务执行部分会提交给底层执行，这就是 Node.js 为什么会有如此高性能的原因。

1.3 Node.js 的应用场景

1.3.1 Node.js 优缺点

在介绍应用场景之前，需要了解一下 Node.js 的利弊。只有在合适的场景下使用 Node.js，才能达到高性能。

Node.js 优点：

- 事件驱动、异步编程，在 I/O 密集型场景下有着极高的性能。
- 轻量高效，资源占用率低。
- 使用 JavaScript（JS）作为应用层语言，语言门槛低，对于拥有 JS 基础的人来说，几乎没有门槛。

Node.js 缺点：

- 单进程，一旦 JS 线程出现未处理的错误，进程会退出，服务会终止。
- 单线程（特指 JS 线程），一旦 JS 线程上出现耗时的 CPU 计算（加解密之类的计算），JS

线程将出现阻塞，会拖慢事件轮询。

1.3.2 应用场景

由于较低的语言门槛以及强大的 I/O 处理能力，大致有以下场景是 Node.js 能够胜任的。

1. Restful API

这是 Node.js 最理想的应用场景，可以处理数万条连接，本身没有太多的逻辑，只需要调用请求 API、组织数据进行返回即可。它本质上只是从某个数据库或缓存中查找一些值并将它们组成一个响应。由于响应是少量的文本，入站请求也是少量的文本，因此流量不高，甚至一台机器也可以处理最繁忙的 API 需求。社区有 Koa、Express、hapi 等框架提供该功能。

2. 实时 WebSocket 应用

大量用户同时在线，互相收发数据，但是几乎不需要对数据进行处理，Node.js 只需要接收数据然后转发。社区有著名的 socket.io 库来提供 WebSocket 功能。

3. 前端工具链

由于采用的语言是 JavaScript，并且拥有大量的第三方模块，在前端工程师手里可以开发一整套前端工具链，包括压缩、混淆、模块化等。比如业界著名的 Webpack。

4. 桌面开发

基于开源的 Chromium 和 Node.js，开发者可以通过 HTML/CSS/JS 来构建桌面端应用程序，业界著名的有 Electron 和 node-webkit。

1.4 本章小结

Node.js 并不是银弹（Silver Bullet），只能解决特定场景下的问题。Node.js 比较有意思的一点是，很多进入 Node.js 世界的是客户端 JavaScript 程序员，此外还有一些使用其他语言进行服务端开发的程序员。不管你是做什么的，我们都希望你能够知道 Node.js 到底适不适合你，适合你的哪些业务。

回顾一下本章所学：

- Node.js 是事件触发和非阻塞的。
- Node.js 专为 I/O 密集型应用而设计。
- Node.js 的模块架构和运行原理。

第 2 章将介绍如何在 Windows/Linux/macOS 系统下安装 Node.js 及其开发环境。

第 2 章

搭建 Node.js 开发环境

本章内容

- Node.js 的安装
- 使用 VSCode 搭建 Node.js 开发环境
- 编写 HTTP 服务器

2.1 安装 Node.js

为了保证下载速度，本文所用的下载地址为淘宝 NPM 镜像源中的 Node.js 二进制镜像。

本书写作时最新的 Node.js 稳定版的版本号为 12.12.0，各操作系统采用的安装都以该版本作为示例。

一般来说，按照书中的步骤来安装不会出现问题，如果出现问题，请把截图反馈给作者公众号或者 GitHub。

2.1.1 Windows 上安装 Node.js

32 位下载地址：https://npm.taobao.org/mirrors/node/latest-v12.x/node-v12.12.0-x86.msi。
64 位下载地址：https://npm.taobao.org/mirrors/node/latest-v12.x/node-v12.12.0-x64.msi。
大家可以根据系统位数来选择对应的安装包进行下载与安装。

2.1.2 Linux 安装 Node.js

Node.js 官方发布的 Linux 通用二进制文件,支持主流的 Linux 发行版(Ubuntu、CentOS 等等)。新版本的 Node.js 发行版不再提供 32 位版本。

64 位下载地址:https://npm.taobao.org/mirrors/node/latest-v12.x/node-v12.0.0-linux-x64.tar.gz。

推荐使用 /opt 目录来部署已经编译好的二进制包,安装步骤如下:

```
# 下载文件
wget https://npm.taobao.org/mirrors/node/latest-v12.x/node-v12.12.0-linux-x64.tar.gz
# 解压并移动到指定目录
mv node-v12.12.0-linux-x64.tar.gz /opt
tar xf node-v12.12.0-linux-x64.tar.gz
mv node-v12.12.0-linux-x64 /opt/nodejs
# 编辑 ~/.bashrc 文件,添加 Node.js 到 PATH 环境变量中
echo 'PATH=/opt/nodejs/bin:$PATH' >> ~/.bashrc
# 更新环境变量
source ~/.bashrc
# 测试安装结果
node -v
npm -v
# 设置 NPM 包镜像源为 taobao
npm config set registry https://registry.npm.taobao.org
```

2.1.3 Ubuntu 安装 Node.js

Ubuntu 官方仓库中提供了 Node.js 和 NPM 包,我们可以直接使用 apt-get 来进行安装。安装步骤如下:

```
sudo apt-get install nodejs npm
# 测试安装结果
node -v
npm -v
# 设置 NPM 包镜像源为 taobao
npm config set registry https://registry.npm.taobao.org
```

2.1.4 CentOS 安装 Node.js

CentOS 官方仓库中也提供了 Node.js 和 NPM 包,但是版本一般比较旧,建议使用上节中 Linux 安装 Node.js 的方式进行安装。安装步骤如下:

```
yum install nodejs -y
# 测试安装结果
node -v
npm -v
```

```
# 设置 NPM 包镜像源为 taobao
npm config set registry https://registry.npm.taobao.org
```

2.1.5 macOS 安装 Node.js

1. pkg 方式

Node.js 官方提供了适用于 macOS 系统的 pkg 安装包，我们直接下载进行安装即可。

64 位下载地址：https://npm.taobao.org/mirrors/node/latest-v12.x/node-v12.12.0.pkg。

安装完毕后需要设置以下 NPM 包的镜像源为淘宝：

```
npm config set registry https://registry.npm.taobao.org
```

2. Homebrew 方式

如果你的计算机中安装了 Homebrew 环境，直接使用以下命令安装即可：

```
brew install node
# 设置 NPM 包镜像源为 taobao
npm config set registry https://registry.npm.taobao.org
```

2.2 安装 VSCode 编辑器

Visual Studio Code（简称 VSCode）是由微软开发的、同时支持 Windows/Linux/macOS 操作系统的开源编辑器。它支持测试功能，并且内置了 git 功能，提供了丰富的语言支持与常用编程工具。

本节将介绍如何在本地计算机上搭建 VSCode 开发环境，以下步骤使用 macOS 版本作为示例，其他操作系统类似。

（1）打开官方网站 https://code.visualstudio.com/，单击蓝色按钮下载即可。

（2）新版本的 VSCode 不再内置中文语言包，需要安装语言包扩展。安装 VSCode 后打开 VSCode 编辑器，在扩展窗口中搜索"Chinese"，安装第一个即可，如图 2-1 所示。

图 2-1

（3）VSCode 内置 Typescript 支持，所以我们开发的时候可以选择使用 JavaScript 语言开发,同时导入 Node.js 的 Typescript definitions 声明文件(类似于 C 语言的头文件),可以获得代码自动提示的功能。

至此,基于 VSCode 的 Node.js 开发环境已经搭建完毕,如果你遇到疑问或者有搭建上的问题,可以在作者公众号或者 GitHub 上进行咨询。

2.3 编写 HTTP 服务器

在编写第一个 Node.js 应用前,先了解一下 Node.js 应用(本节指服务端的应用)是由哪些部分组成的:

(1)通过 require 或 import 导入依赖模块。
(2)创建服务器并设置好事件回调。
(3)启动服务器。

使用 VSCode 编辑器新建 app.js 文件并写入以下代码:

```
const http = require('http');    // 导入 http 模块

const hostname = '127.0.0.1';    // HTTP 服务器监听地址
const port = 3000;               // HTTP 服务器监听端口

const server = http.createServer((req, res) => {  // 事件回调
  res.statusCode = 200;          // 设置 HTTP 响应状态码
  res.setHeader('Content-Type', 'text/plain');   // 设置响应内容格式
  res.end('Hello World\n');      // 输出内容并结束此次响应
});

server.listen(port, hostname, () => {            // 运行服务器
  console.log(`Server running at http://${hostname}:${port}/`);
});
```

单击如图 2-2 所示的菜单选项打开 VSCode 内置的终端:

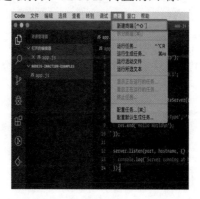

图 2-2

输入 node app.js 即可启动 HTTP 服务器，如果出现启动失败，一般是端口被占用的原因，只要更换监听的端口即可，如图 2-3 所示。

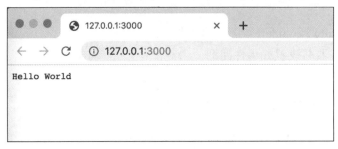

图 2-3

打开浏览器访问 http://127.0.0.1:3000，即可看到如图 2-4 所示的界面。

图 2-4

2.4 本章小结

Node.js 的服务器开发过程比传统的 Web 服务器开发要简洁很多，底层为了处理大量关于 HTTP 协议的实现细节，只提供了基本且必要的 API 给 JS 层，这样我们就可以专注业务的开发，而不用消耗时间和精力来处理协议层。

回顾一下本章所学：

- 主流操作系统下 Node.js 的安装。
- VSCode 开发环境的搭建流程。
- Node.js 应用的开发流程。

有的读者看到本章的 HTTP 模块、事件的编写方式可能会存在疑问，没关系，有疑问是正常的，因为 Node.js 的大门才刚刚开启。

第 3 章将重点学习 Node.js 编程基础。

第 3 章

Node.js 编程基础

本章内容

- NPM 包管理器介绍
- package.json 文件介绍
- Node.js 的模块系统
- Node.js 的异步编程方式
- Node.js 常用模块

3.1　NPM 包管理器介绍

Node.js 中包管理器是用来管理 Node.js 软件包的工具，类似于 Java 的 Maven 或者 PHP 的 Composer。

NPM（Node.js Package Manager）是 Node.js 默认的包管理工具，能够解决 Node.js 开发和部署中软件包依赖的问题。常见的使用场景有以下几种：

- 从 NPM 服务器下载别人编写的第三方包到本地进行使用。
- 将自己编写的软件包上传到 NPM 服务器供他人使用。

3.1.1　更换 NPM 镜像源

在国内访问 NPM 默认的中央仓库速度比较慢，可以更换为淘宝提供的 NPM 镜像源以加快软件包的安装。

在终端按需执行以下命令:

```
# 设置镜像源为淘宝
npm config set registry https://registry.npm.taobao.org
# 还原官方镜像源
npm config delete registry
```

使用淘宝镜像之后无法使用 publish 和 unpublish 命令,如果需要发布软件包和撤销发布的软件包,则需要还原为官方镜像。

3.1.2 初始化项目

在项目目录下执行 npm init 命令,依照提示输入问题的答案之后即可创建一个标准的 npm 模块,同时会生成一个 package.json 文件,其中记录了当前的模块名、版本、依赖等信息。

3.1.3 使用 npm 命令安装模块

安装 Node.js 模块的 npm 命令语法如下:

```
npm install <模块名称>
```

如果需要使用常用的 Koa 框架进行开发,则可以使用如下命令进行安装:

```
npm install koa
```

该命令执行完毕之后,Koa 模块就会出现在当前目录下的 node_modules 文件夹中,使用如下代码即可使用该模块:

```
const koa = require('koa');
```

3.1.4 本地安装与全局安装

npm 的模块安装分为全局安装和本地安装,默认为本地安装,如果需要全局安装,则要使用 -g 参数:

```
npm install express-generator -g
```

1. 本地安装
- 第三方模块将被安装到当前目录的 node_modules 下(如果没有该目录则会自动新建)。
- 通过 require('模块名') 即可导入本地模块。

2. 全局安装
- 第三方模块将被安装到 /usr/local/lib/node_modules 目录或者安装 Node.js 的目录。
- 可以直接在命令行使用。
- 不可以使用 require('模块名') 来引用。

3.1.5 生产依赖和开发依赖

有一些软件包是开发环境和生产环境都需要的,而有一些只在开发环境使用,比如测试框架。

1. 开发依赖

使用--save-dev 选项即可将软件包安装为开发依赖,依赖信息将被保存到 package.json 的 devDependencies 中。

```
npm install eslint --save-dev
```

2. 生产依赖

使用--save 选项即可将软件包安装为生产依赖,开发环境和生产环境都需要使用,依赖信息将被保存在 package.json 的 dependencies 中。

```
npm install koa --save
```

3.1.6 其他 npm 命令

其他 npm 命令如表 3-1 所示。

表 3-1 其他 npm 命令

命令	说明
npm uninstall <模块名> [-g]	卸载本地/全局模块
npm list [-g]	查看已安装的本地/全局模块
npm update <模块名> [-g]	更新本地/全局模块
npm search <模块名>	搜索模块
npm publish	发布本地模块到 NPM 仓库
npm unpublish <模块名>@<版本号>	撤销指定版本的模块发布
npm config get <config_name>	读取配置
npm config set <config_name>	设置配置
npm config delete <config_name>	删除配置
npm run <命令>	执行 package.json 中 scripts 定义的命令

3.2 Yarn 包管理器介绍

Yarn 是 Facebook 发布的一款 NPM 包管理工具，解决了 NPM 包下载速度慢、每次都要全量下载、版本号依赖混乱等问题。

3.2.1 安装 Yarn

Yarn 是命令行工具，需要安装到全局模块中，在终端执行以下命令即可：

```
npm install yarn -g
```

3.2.2 Yarn 常用命令

Yarn 常用命令如表 3-2 所示。

表 3-2 Yarn 常用命令

命令	说明
yarn [global] remove <模块名>	卸载本地/全局模块
Yarn [global] upgrade <模块名>	更新本地/全局模块
yarn [global] add <模块名>	安装本地/全局模块
yarn config get <config_name>	读取配置
yarn config set <config_name>	设置配置
yarn config delete <config_name>	删除配置
yarn run <命令>	执行 package.json 中 scripts 定义的命令

笔者推荐各位读者使用 Yarn 来进行软件包依赖的管理和下载。

3.3 解读 package.json 文件

package.json 是 Node.js 软件包的元数据（Meta）描述文件，一般由 npm/yarn 命令创建，不需要手动维护。一个典型的 package.json 如下：

```json
{
  "name": "example",
  "version": "1.0.0",
  "description": "example Node.js package",
  "main": "index.js",
  "scripts": {
    "test": "NODE_ENV=test mocha -t 25000"
  },
  "repository": {
    "type": "git",
    "url": "git://github.com/xxx/xxx.git"
  },
  "bugs": {
    "url": "https://github.com/xxx/issues"
  },
  "keywords": [
    "storage"
  ],
  "author": "xialeistudio@gmail.com",
  "contributors": [
  ],
  "dependencies": {
    "urllib": "^2.34.1"
  },
  "devDependencies": {
    "mocha": "^6.2.1",
    "should": "^13.2.3"
  },
  "license": "MIT"
}
```

3.3.1　package.json 字段说明

package.json 字段说明如表 3-3 所示。

表 3-3　package.json 字段及其说明

字段名称	字段说明
name	软件包名称
version	软件包版本
description	软件包简介
main	软件包入口
scripts	项目命令
repository	仓库地址
bugs	bug 反馈页面

(续表)

字段名称	字段说明
keywords	软件包关键字
author	软件包作者
contributors	软件包贡献者
dependencies	生产依赖
devDependencies	开发依赖
license	开源协议

3.3.2 版本号说明

NPM 包使用语义化的版本号来管理代码，版本号格式为 X.Y.Z，分别代表主版本号、次版本号和补丁版本号。当代码有修改时，需要按照以下规则执行版本号的变更：

- 只是修复 bug，更新 Z 位。
- 只是新增功能，但是向下兼容（旧 API 不受影响），更新 Y 位。
- 向下不兼容的改动，更新 X 位。

3.3.3 常见的版本号限定符

在 package.json 中会见到类似^0.1.0,~0.1.0,0.1.0 和>=0.1.0 之类具有不同限定符的版本号，为了避免混淆，这里做一下说明。

- ^0.1.0：支持 0.1.0～1.0.0（不含）之内的所有版本。
- ~0.1.0：支持 0.1.0～0.2.0（不含）之内的所有版本。
- 0.1.0：只能使用 0.1.0 版本。
- >= 0.1.0：支持大于等于 0.1.0 之后的所有版本。
- *：任意版本。

3.4　Node.js 的模块系统

Node.js 应用由模块组成，采用 CommonJS 模块规范。

每个 JS 文件就是一个模块，有独立的作用域。在一个文件中定义的变量、函数、类在其他文件都不可见。

```
// example.js
function sum(a, b) {
```

```
  return a + b;
}
```

上述代码中 sum 函数只有 example.js 中能调用，其他文件则不可以调用。

3.4.1 module 和 exports

1. 基本用法

CommonJS 规范规定，每个模块内部的变量 module 代表当前模块。

module 是一个对象，module.exports 是对外的接口。加载模块实际上是读取该模块的 module.exports 属性。

module.exports 也是一个对象，所有需要导出的变量、函数、类都需要挂载到该对象上才能实现导出。

```
// example.js
function sum(a, b) {
  return a + b;
}

module.exports.sum = sum;
```

为了方便起见，Node.js 为每个模块提供了一个 exports 变量，在同一个模块中，module.exports 和 exports 是恒等的（类型和值都相等）。因此在实际开发中，建议通过 exports.xxx 的形式导出变量、函数、类。

不建议更改 exports 的指向，否则模块将不能正常导出。

下列代码无法导出 sum 函数，因为 exports 由于重新赋值导致指向被更改。

```
// example.js
exports.sum = function(a, b) {
  return a + b;
}
exports = 'Hello World';
```

如果模块只需要导出一个变量、函数、类，只能对 module.exports 进行赋值，对 exports 进行赋值达不到如期作用。

上面讲到 module.exports 是恒等于 exports 的，为什么对 exports 进行赋值达不到如期作用呢？

```
// a.js
exports = 'hello world';
// b.js
module.exports = 'hello world';
// c.js
const hello = require('./a');
console.log(hello); // {}

const hello2 = require('./b');
```

```
console.log(hello2); // hello world
```

2. 原理解读

exports 只是一个别名，类似于下面的代码：

```
    var exports = module.exports;
```

当不对 exports 重新赋值时，exports 指向不变，exports.xxx 也会如期地添加到 module.exports 中。当对 exports 重新赋值时，exports 和 module.exports 关联就不存在了，修改 exports 不会对 module.exports 产生作用。

3.4.2　require

1. 基本用法

CommonJS 规定 require 用于加载模块文件。

require 读取并执行一个 JS 模块，然后返回该模块的 exports 对象。如果模块未找到，则会抛出错误。

```
// math.js
exports.sum = function(a, b) {
  return a + b;
}

// index.js
const math = require('./math');
math.sum(1, 1);
```

2. 加载规则

加载模块时模块扩展名为.js。也就是说下列代码是一样的：

```
const math = require('./math');
const math2 = require('./math2');
```

根据传入的参数，require 会有不同的规则（以下规则无先后顺序）：

（1）如果参数以'/'开头，则表示需要加载的是一个绝对路径的 JS 文件。如 require('/home/xialei/math')将加载/home/xialei/math.js。

（2）如果参数字符串是以'./'开头，则表示需要加载一个相对路径的模块文件。如 require('./math')将加载位于当前模块同目录下的 math.js 文件。

（3）如果参数不以'./'或'/'开头，则需要加载的是核心模块或者当前工作目录中 node_modules 下的模块。

（4）如果没有找到指定的模块文件，Node.js 会尝试自动添加.js、.json、.node（编译后的二进制模块）后再去搜索。

（5）如果传入的参数解析之后是一个目录，Node.js 会自动读取该目录下的 package.json 文件，根据 main 字段来加载真正的入口文件。如果该目录下没有 package.json 文件，则尝试加载 index.js 或 index.node。

3.4.3 开发一个自定义模块

我们将开发一个与时间操作相关的函数模块来巩固本节所学，示例代码如下。
date.js:

```
exports.formatTime = function (timestamp) {
    const duration = parseInt(Date.now() / 1000, 10) - timestamp;
    if (duration < 600) {              // 不足10分钟
        return '刚刚';
    }
    if (duration < 3600) {             // 不足1小时
        return '1小时内';
    }
    if (duration < 3600 * 3) {         // 不足3小时
        return '3小时内';
    }
    if (duration < 3600 * 24) {        // 不足24小时
        return '今天';
    }
    if (duration < 3600 * 24 * 2) {    // 不足两天
        return '1天前';
    }
    return new Date(timestamp * 1000).toString();
}
```

index.js:

```
const date = require('./date');

const now = parseInt(Date.now() / 1000, 10);

console.log(date.formatTime(now - 60));
console.log(date.formatTime(now - 600));
console.log(date.formatTime(now - 5400));
console.log(date.formatTime(now - 3600 * 23));
console.log(date.formatTime(now - 3600 * 24));
console.log(date.formatTime(now - 3600 * 24 * 3));
```
在终端输入以下命令执行 index.js:
node index.js

输出如下：

```
刚刚
1 小时内
3 小时内
今天
1 天前
Fri Oct 23 2019 18:07:45 GMT+0800 (中国标准时间)
```

3.5　Node.js 的异步编程风格

在本书的第 1 章介绍过 Node.js 是一个异步运行环境，异步意味着调用函数后结果不是立即返回，而是在未来的时刻再通知给调用者。Node.js 使用最广泛的异步编程风格是基于回调函数来实现的。

截止本书出版时，Node.js 可以使用以下几种异步编程风格：

- 回调函数
- Promise
- async/await

回调函数是最早出现的，也是最烦琐的，实际应用中不推荐以回调函数的形式进行异步调用。原因是多个异步操作有顺序依赖时，会产生如下代码：

```
func1.success(function() {
  func2.success(function() {
    func3.success(function() {

    });
  });
});
```

3.5.1　回调函数

Node.js 异步编程是通过回调函数来实现的，但不能说使用回调函数就异步了。

如下代码是基于回调函数的实现，但不是异步的：

```
function test(callback) {
  callback(1);
}

test(function(data) {
  console.log(data);
});
```

回调函数在完成任务后就会被调用，Node.js 使用了大量回调函数，几乎所有的 API 都支持回调函数。

由于调用接口存在成功或失败的情况，而基于回调函数的编程无法使用标准 JS 中的抛出错误和捕获错误的方法。因此只能将错误对象作为回调参数来调用回调函数。

Node.js 中回调函数的风格是统一的，这样给我们编程带来了很大的方便。异步函数的签名如下：

```
func(param..., callback(Error, data))
```

- param 调用 API 的参数。如读取文件时传递的文件路径，支持多个参数。
- callback(Error, data...) 回调函数。Node.js 中回调函数的第一个参数永远是 Error 对象，之后才是调用成功的结果，如果没有出错，第一个参数为 null。

例如，我们读取位于桌面的 data.txt 文件：

```
const fs = require('fs');   // fs 是 Node.js 核心模块
fs.readFile('~/Desktop/data.txt', function(err, data) {
  if(err) {
    console.log('读取出错', err);
    return;
  }
  console.log(data);
});
```

3.5.2 Promise

1. 基本知识

Promise 对象用于表示一个异步操作的最终完成（或失败）及其结果值。

一个 Promise 有以下几种状态：

- pending：初始状态
- fulfilled：操作成功
- rejected：操作失败

Promise 只会从 pending 转换为 fulfilled 或者 rejected，整个转换只发生一次。

Promise 构造函数接收一个执行函数，该函数接收 resolve 和 reject 两个回调函数，当执行函数运行成功时需调用 resolve，执行错误时需调用 reject。

Promise 构造函数的签名如下：

```
function Promise(function(resolve, reject): Promise {
    // 原来的异步逻辑
});
```

一旦 Promise 发生状态变化，就会触发 then 方法，then 方法签名如下：

```
Promise.prototype.then = function(onFulfilled[, onRejected]): Promise
```

- onFulfilled Promise：执行成功时回调。
- onRejected Promise：执行出错时回调，该参数是可选的。
- then 方法：返回一个新的 Promise 对象，因此 Promise 支持链式调用。

由于 then 的第二个参数 onRejected 参数是可选的，因此 Promise 的原型上提供了 catch 方法来捕获异步错误，catch 方法签名如下：

```
Promise.prototype.catch = function(onRejected): Promise
```

- onRejected Promise：执行出错时回调。
- catch 方法：返回一个新的 Promise 对象。

2. 基本使用

Promise 是为了解决异步编程问题而出现的，因此可以基于 Promise 来优化上文中读取文件的例子：

```
function readFileAsync(path) {
  return new Promise(function(resolve, reject) {    // 实例化 Promise
    fs.readFile(path, function(err, data) {         // 执行原定的异步逻辑
      if(err) {
        reject(err);
        return;
      }
      resolve(data);
    });
  });
}

readFileAsync('~/Desktop/data.txt')
 .then(function(data) {
  console.log(data);
}).catch(function(err) {
  console.error(err);
});
```

3. 链式调用

Promise 的 then 或 catch 回调函数的返回值会作为下一个 then/catch 的输入参数，因此可以通过链式 Promise 来扁平化嵌套的回调函数。

```
// callback.js 回调风格
const fs = require('fs');

fs.readFile('~/Desktop/data.txt', function(err, data) {
  if(err) {
    console.log('读取 data.txt 出错', err);
    return;
  }
  console.log('data.txt 读取成功', data);

  fs.readFile('~/Desktop/data2.txt', function(err, data2) {
    if(err) {
      console.log('读取 data2.txt 出错', err);
      return;
    }
    console.log('data2.txt 读取成功', data2);
  });
});
```

如果需要依次读取两个文件，那么就需要嵌套一层回调函数。如果依赖的异步操作越多，响应的嵌套层级也会越大，给代码的可读性和可维护性带来困难。

```javascript
// promise.js Promise 风格
const fs = require('fs');
function readFileAsync(path) {
  return new Promise(function(resolve,reject) {
    fs.readFile(path, function(err, data) {
      if(err) {
        reject(err);
        return;
      }
      resolve(data);
    });
  });
}
readFileAsync('~/Desktop/data.txt')
.then(function(data) {
  console.log('读取 data.txt 成功', data);
  return readFileAsync('~/Desktop/data2.txt');        // 返回 Promise
})
.then(function(data2) {
  console.log('读取 data2.txt 成功', data2);
})
.catch(function(err) {
  console.log('读取失败', err)
});
```

多个 Promise 链式调用时一旦有一个出错，整个调用链就会终止，然后回调 catch 函数。通过链式调用，困扰多年的 Node.js 回调嵌套问题终于得到了第一次解决。

4. 其他操作

`Promise.resolve(value)`

返回一个状态由 value 决定的 Promise 对象。value 有以下几种取值：

- Promise。value 本身是 Promise 的情况下，返回的 Promise 值由 value 这个 Promise 决定。
- 基本类型/空/或不带 then 方法的对象。返回的 Promise 值为 value，状态为 fulfilled。

`Promise.reject(reason)`

返回一个状态为失败的 Promise，通常情况下会传递 Error 对象作为 reason。

`Promise.all(promises)`

接收一个 Promise 数组，返回一个新的 Promise 对象。

- Promise 数组中所有 Promise 都成功执行的情况下，返回的 Promise 最终会触发成功。

- Promise 数组中只要有一个 Promise 执行失败,返回的 Promise 最终会执行失败。

```
Promise.race(promises)
```

接收一个 Promise 数组,返回一个新的 Promise 对象。

当 Promise 数组中任意一个 Promise 执行成功或失败,返回的 Promise 则立即成功或失败。

3.5.3 async/await

async 和 await 关键字是 ES2017 中新添加的关键字,本质上是 Promise 的语法糖,使得能够像同步代码一样编写异步代码。

async/await 本质是语法糖,因此尽管编程风格与同步类似,但是不会阻塞 JS 线程。

下面使用 Promise 小节中依次读取两个文件的需求为例:

```
// async.js
const fs = require('fs');

function readFileAsync(path) {
  return new Promise(function(resolve,reject) {
    fs.readFile(path, function(err, data) {
      if(err) {
        reject(err);
        return;
      }
      resolve(data);
    });
  });
}

// 依次读取文件
async function readFiles() {
  try {
    const data1 = await readFileAsync('~/Desktop/data.txt');
    const data2 = await readFileAsync('~/Desktop/data2.txt');
    console.log('文件 1 内容', data1);
    console.log('文件 2 内容', data2);
  } catch(e) {
    console.error('读取失败', e);
  }
}

readFiles();
```

可以看到 readFiles 函数中的代码跟同步编程风格一致(忽略 async/await 关键字)。

针对 Promise 调用链太长的问题,async/await 提供了一个优美的解决方案,确实让异步编程变得简单了。

基本语法

（1）async

async 只能放在函数声明之前，支持普通函数、箭头函数和类函数。被修饰的函数不管返回什么值，最终都会返回 Promise。

- 函数返回基本值/空/或不带 then 方法的对象时，Promise 的结果为该值，状态 fulfilled。
- 函数抛出错误时，Promise 状态为 rejected，reason 为抛出的错误对象。
- 函数本身返回一个 Promise 时，最终的 Promise 结果为该 Promise 的结果。

```
// 普通函数
async function doSomething(a, b) {
  return a + b;
}
// 箭头函数
const doSomething2 = async (a, b) => {
  return a + b;
};
// 类函数
class Demo {
 async test(a, b) {
   return a + b;
 }
}

const data = doSomething(1, 2); // [object Promise]
doSomething(1, 2).then(function(result) {
  console.log(result); // 3
});
```

由于被 async 修饰的函数最终都会返回 Promise，因此需要使用 then 才可以获得 Promise 的执行结果，直接调用函数只能得到一个 Promise。

（2）await

await 只能在被 async 修饰的函数内部调用，await 可以放在任何返回 Promise 的函数前，Promise 执行成功的情况下，await 语句将返回 Promise 的成功值，Promise 执行错误的情况下，await 语句将抛出错误，通过 try/catch 捕获即可。

示例: 包装 XMLHttpRequest

```
function ajaxGet(url, timeout) {
    return new Promise(function (resolve, reject) {
        const xhr = new XMLHttpRequest();
        xhr.open('GET', url, true);
        xhr.timeout = timeout;

        xhr.onload = function () {
            resolve(xhr.responseText);
        }
        xhr.onerror = function (e) {
```

```
            reject(new Error('请求失败'));
        };
        xhr.ontimeout = function () {
            reject(new Error('timeout'));
        };

        xhr.send(null);
    });
}
async function getData() {
    try {
        const data = await ajaxGet('xx://a.json', 1000); // 1 秒超时
        console.log(data);
    } catch (e) {
        console.error(e);
    }
}
getData();
```

几乎所有 callback 类型的异步函数都可以包装为 Promise（特殊情况就是 callback 有多个返回值的情况，Promise 不能直接处理，但是可以将多个返回值包装为一个数组）。

3.6　Node.js 常用核心模块

学习 Node.js 必须掌握其核心模块，就像学习 JavaScript 必须掌握函数、对象、数据类型、DOM、BOM 等知识一样。

Node.js 的核心模块很多，这里介绍几个比较常用的：events、fs、stream、http 模块。

3.6.1　events 模块

事件驱动、非阻塞 IO 是 Node.js 的特点，所以事件模块是非常重要的模块。Node.js 绝大多数模块都继承了 events。

基于事件的编程是典型的发布-订阅模式的实现，有效地解耦了发布者和订阅者，发布者只需要关心事件的触发，不需要关心触发之后的逻辑，而订阅者只关注订阅，不需要关注事件是由谁来触发。

1．基本使用

使用事件模块有以下几个步骤：

（1）实例化事件监听器实例。

（2）注册事件。

(3）触发事件。

```
const EventEmitter = require('events');
// 实例化事件监听器
const emitter = new EventEmitter();

// 注册事件
emitter.on('click', function (param1) {
    console.log('触发了点击', param1);
});

// 触发事件
emitter.emit('click', 'demo');
// 再次触发
emitter.emit('click', 'demo');
```

以上例程输出如下：

```
触发了点击 demo
触发了点击 demo
```

2. 一次性监听

Node.js 中可以对同一个事件进行多次监听，也可以多次触发同一事件。如果需要在事件触发监听函数一次之后不再继续触发，可以调用 once() 来进行一次性监听。

```
const EventEmitter = require('events');

const emitter = new EventEmitter();

emitter.once('click', function (param1) {
    console.log('触发了点击', param1);
});

emitter.emit('click', 'demo');
emitter.emit('click', 'demo');
```

以上例程输出如下：

```
触发了点击 demo
```

调用 once 函数监听的事件不管触发了多少次事件，都只会执行一次回调函数。

3.6.2 fs 模块

fs 是 File System 的缩写，也就是 Node.js 用来操作系统文件的模块。

1. 读取文件

调用 fs.readFile() 函数即可读取文件，函数签名如下：

```
fs.readFile(path[, options], callback)
```

- path：文件路径。
- options：选项。
- callback：回调函数。

```
const fs = require('fs');

fs.readFile('./a.js', 'utf8', (err, data) => {
   if (err) {
       console.log('读取失败', err);
       return;
   }
   console.log('读取成功', data);
});
```

2. 写入文件

调用 fs.writeFile() 可以写入文件，函数签名如下：

```
fs.writeFile(path, content[, options], callback)
```

- path：文件路径。
- content：文件内容。
- options：选项。
- callback：回调函数。

```
const fs = require('fs');

fs.writeFile('./app.log', (new Date()).toString(), { encoding: 'utf8' }, (err) =>
   {
   if (err) {
       console.log('写入失败', err);
       return;
   }
   console.log('写入成功');
});
```

3. 追加内容

如果文件存在时不需要覆盖，可以调用 fs.appendFile() 来追加文件内容，函数签名如下：

```
fs.appendFile(path, content[, options], callback)
```

- path：文件路径。
- content：文件内容。
- options：选项。
- callback：回调函数。

```
const fs = require('fs');

fs.appendFile('./app.log', (new Date()).toString(), { encoding: 'utf8' }, (err)
   => {
   if (err) {
```

```
        console.log('写入失败', err);
        return;
    }
    console.log('写入成功');
});
```

4. 删除文件

调用 fs.unlink()可以删除文件，函数签名如下：

```
fs.unlink(path, callback)
```

- path：文件路径。
- callback：回调函数。

```
const fs = require('fs');

fs.unlink('./aaa', (err) => {
    if (err) {
        console.log('删除失败', err);
        return;
    }
    console.log('删除成功');
});
```

本函数不能删除文件夹，需要调用 fs.rmdir()才可以删除文件夹。

5. 创建文件夹

调用 fs.mkdir()可以创建文件夹，函数签名如下：

```
fs.mkdir(path[, options], callback)
```

- path：文件夹路径。
- options：选项。
- callback：回调函数。

```
const fs = require('fs');

fs.mkdir('aaa/bbb', { recursive: true }, (err) => { // recursive 选项在创建多级目录
    时使用
    if (err) {
        console.log('创建失败', err);
        return;
    }
    console.log('创建成功');
});
```

6. 读取文件夹内容

调用 fs.readdir()可以读取文件夹内容，函数签名如下：

```
fs.readdir(path[, options], callback)
```

- path:文件夹路径。
- options:选项。
- callback:回调函数。

```
const fs = require('fs');
fs.readdir('.', (err, files) => {
    if (err) {
        console.log('遍历失败', err);
        return;
    }
    console.log(files);
});
```

7. 删除文件夹

调用 fs.rmdir() 可以删除文件夹,函数签名如下:

```
fs.rmdir(path, callback)
```

- path:文件夹路径。
- callback:回调函数。

```
const fs = require('fs');
fs.rmdir('./aaa', (err) => {
    if (err) {
        console.log('删除失败', err);
        return;
    }
    console.log('删除成功');
});
```

3.6.3 stream 接口

stream 是 Node.js 的一个抽象接口,很多对象都实现了这个接口。比如请求 HTTP 接口时的 request 对象就是一个 stream。

stream 是 EventEmitter 的实例。常用的事件有:

- data:当有数据可读时触发。
- end:当没有更多数据可读时触发。
- error:在读取或写入时发生错误时触发。
- finish:在所有数据写入系统底层时触发。

Node.js 中 stream 有 4 种类型:

- Readable:可读。
- Writable:可写。
- Duplex:可读可写。

- Transform：数据转换。

1. 读取流

读取流都可以通过监听 data 和 end 事件来接收数据。比如文件读取流、HTTP 请求读取流都可以用这个方法来读取，不需要关心读取流的类型。

通过流的方式读取文件有个很大的优点：内存占用低，可以以小块的形式处理数据。而直接调用 fs.readFile()之类的函数，如果文件过大，可能会撑爆内存。

```javascript
// a.js
const fs = require('fs');

let data = '';
const stream = fs.createReadStream('./a.js', { encoding: 'utf8' });

stream.on('data', (chunk) => {      // chunk 是本次读取到的数据
    data += chunk;
});

stream.on('end', () => {            // 读取文件完毕
    console.log(data);
});

stream.on('error', (err) => {
    console.error('读取错误', err);
});

console.log('执行完毕');
```

2. 写入流

通过调用写入流的 write()方法来写入数据，调用写入流的 end()来完成写入。当数据写入完毕时会触发 finish 事件。

```javascript
// a.js
const fs = require('fs');

const readStream = fs.createReadStream('./a.js', { encoding: 'utf8' });
const writeStream = fs.createWriteStream('./b.js', { encoding: 'utf8' });

readStream.on('data', (chunk) => {  // 读取到小块数据就写入，内存占用低
    writeStream.write(chunk);
});

readStream.on('end', () => {        // 文件读取完毕
    writeStream.end();              // 写入完成
    console.log('读取完成');
});

writeStream.on('finish', () => {
    console.log('写入完成');
```

```js
});
readStream.on('error', (err) => {
    console.log('读取错误', err);
});
writeStream.on('error', (err) => {
    console.log('写入错误', err);
});
```

3. 管道流

管道提供了一个读取流到写入流的机制。我们通常用管道从一个流中获取数据并将数据传递到另外一个流中。多个管道可以串行处理，数据会依次流经每个管道。

实际上在上面的写入流示例中，已经体现了管道流的思想。只不过通过调用读取流的 pipe() 方法可以简化该步骤。

```js
// a.js
const fs = require('fs');

// 创建读取流
const readStream = fs.createReadStream('./a.js', { encoding: 'utf8' });

// 创建写入流
const writeStream = fs.createWriteStream('./b.js', { encoding: 'utf8' });

readStream.pipe(writeStream);          // 管道操作

writeStream.on('finish', () => {       // 写入完毕
    console.log('写入完成');
});
```

4. 数据转换流

数据转换流本质还是利用管道进行处理，将读取流通过 pipe() 操作传入转换流，以达到数据转换的目的。

比如我们需要将一个文本文件通过 gzip 压缩后保存到磁盘，可以使用下面的代码。

```js
// a.js
var fs = require("fs");
var zlib = require('zlib');

fs.createReadStream('./a.js')      // 创建读取流
  .pipe(zlib.createGzip())          // 创建 gzip Transform 流
  .pipe(fs.createWriteStream('a.js.gz'));   // 创建写入流

console.log("文件压缩完成。");
```

3.6.4　http 模块

http 模块主要用于创建 HTTP 服务器处理请求、创建 HTTP 客户端发出请求。

1. http 客户端

http 客户端可以向指定的 URL 发出请求，回调一个读取流，可以通过操作读取流进行数据输出或者保存等操作。

下面以保存笔者博客所在的头像文件为例：

```
// 头像链接：https://www.ddhigh.com/images/logo.svg

// 博客是https协议，使用https模块。如果是http协议，则使用http模块，其他代码相同
const https = require('https');
const fs = require('fs');

const req = https.request('https://static.ddhigh.com/blog/2019-09-18-094336.jpg',
    (response) => {
    console.log('响应状态码', response.statusCode);
    response.pipe(fs.createWriteStream('logo.jpg')); // 通过管道流保存图片
});

req.end(); // 发出请求
```

2. http 服务器

http 服务可以说是 Node.js 最为常见的一种服务类型，本节中只简单做个示例。在实际开发中，很少直接使用 Node.js 内置的 http 模块进行开发，一般都是使用 Web 框架。

```
const http = require('http'); // 导入HTTP模块

// 创建HTTP服务器，req为请求对象，resp为响应对象
const server = http.createServer((req, resp) => {
    resp.end(JSON.stringify(req.headers));
});

server.listen(8080, () => console.log('listen on 8080')); // 监听端口
```

执行该 JS 之后会监听本机 8080 端口，通过浏览器访问可以得到如图 3-1 所示的输出。

```
{
    host: "localhost:8080",
    connection: "keep-alive",
    dnt: "1",
    upgrade-insecure-requests: "1",
    user-agent: "Mozilla/5.0 (Macintosh; Intel Mac OS X 10_14_5)
    sec-fetch-user: "?1",
    accept: "text/html,application/xhtml+xml,application/xml;q=0.
    sec-fetch-site: "none",
    sec-fetch-mode: "navigate",
    accept-encoding: "gzip, deflate, br",
    accept-language: "zh-CN,zh;q=0.9,en;q=0.8,zh-TW;q=0.7",
    cookie: "Idea-26226ea6=760672c4-a540-4189-a6d4-
    8d71c8b48f59; grafana_session=20e77e60cad3c900ad12cf8d6dc2c1c
}
```

图 3-1

3.7 本章小结

本章介绍了 Node.js 编程基础，这些内容偏基础一点，主要是为了让初学者了解并熟悉 Node.js 应用的开发流程和核心模块的使用。

回顾一下本章所学：

- Node.js 包管理器 NPM 和 Yarn 基本功能的使用。
- Node.js 的模块系统。
- Node.js 异步编程的三种形式：callback/Promise/async && await。
- Node.js 常用核心模块 events/fs/stream/http。

第二部分　后端的Node.js

Node.js 开发 IO 密集型的后端服务端程序具有得天独厚的语言优势和性能优势。在第二部分，我们将学习主流的 Web 框架和常用组件，涵盖 Express、Koa、MongoDB、MySQL、Redis 等。

第 4 章

最流行的 Web 框架——Express

本章内容

- Express 的基础使用
- Express 的路由功能
- Express 的请求与响应对象
- Express 的中间件
- Express 的错误处理
- Express 的页面渲染与数据输出
- 基于 Express 框架开发留言板系统

4.1 框架简介

Express 是一个简洁而灵活的 Node.js Web 应用框架，提供了一系列强大特性来帮助我们创建各种 Web 应用。使用 Express 可以快速地搭建一个完整功能的网站。

Express 框架的核心特性如下：

- 通过设置中间件来处理 HTTP 请求。
- 通过路由来执行不同的 HTTP 请求操作。
- 通过模板来渲染 HTML 页面。

4.2 快速开始

本节将学习如何基于 Express 框架来开发一个 HTTP 服务器。

4.2.1 初始化项目

新建应用目录，然后进入该目录并将其作为工作目录：

```
mkdir express-example
cd express-example
```

通过 npm 创建一个 package.json 文件：

```
npm init
```

此命令将要求你输出几个参数，例如应用的名称和版本。在本节的内容中，你可以直接按回车键接受默认认值。

接下来安装 Express 并将其保存到 package.json 的依赖列表中：

```
npm install express --save
```

4.2.2 开始编码

新建 app.js，代码如下：

```js
// 导入 express 模块
const express = require('express');
// 创建应用
const app = express();

// 设置路由
app.get('/', (req, resp) => {
    // 输出响应
    resp.json(req.headers);
});

// 开启监听
app.listen(8080, () => {
    console.log('listen on 8080');
});
```

4.2.3 运行应用

在终端执行该 JS：

```
node index.js
listen on 8080
```

使用浏览器访问 http://localhost:8080，结果如下：

```
{
    "host": "localhost:8080",
    "connection": "keep-alive",
    "cache-control": "max-age=0",
    "dnt": "1",
    "upgrade-insecure-requests": "1",
    "user-agent": "Mozilla/5.0 (Macintosh; Intel Mac OS X 10_14_5)
AppleWebKit/537.36 (KHTML, like Gecko) Chrome/78.0.3904.87 Safari/537.36",
    "sec-fetch-user": "?1",
    "accept":
"text/html,application/xhtml+xml,application/xml;q=0.9,image/webp,image/ap
ng,*/*;q=0.8,application/signed-exchange;v=b3",
    "sec-fetch-site": "none",
    "sec-fetch-mode": "navigate",
    "accept-encoding": "gzip, deflate, br",
    "accept-language": "zh-CN,zh;q=0.9,en;q=0.8,zh-TW;q=0.7",
    "if-none-match": "W/\"2c9-A5ngF548rGoZQ5LBRc4RaCa3xh8\""
}
```

4.2.4 小结

本节使用 Express 开发了一个快速入门示例，开发步骤如下：

（1）创建应用实例。
（2）设置路由。
（3）开启监听。

采用 Express 框架的优点如下：

- 支持路由，Node.js 的 http 模块路由功能需要自己开发。
- 支持直接输出 JSON，要让 Node.js 的 http 模块输出 JSON，则需要调用 JSON.stringify()。

4.3 路 由

路由是指应用程序如何根据指定的路由路径和指定的 HTTP 请求方法（GET 和 POST 等）

来处理请求。

在 Express 应用中，每个路由可以有一个或多个处理函数，这些函数会在路由匹配时执行。路由采用以下结构定义：

```
app.METHOD(PATH, HANDLER)
```

- app：应用实例。
- METHOD：小写 get 和 post 等的 HTTP 请求方法。
- PATH：路由路径。
- HANDLER：路由匹配时执行的函数。

4.3.1 路由方法

路由方法是从 HTTP 方法派生的，以下是 GET 和 POST 方法定义路由的示例：

```
app.get('/', (req, resp) => {
  resp.send('GET 请求');
});

app.post('/', (req, resp) => {
  resp.send('POST 请求');
});
```

Express 支持所有的 HTTP 请求方法：

- get
- post
- head
- options
- delete
- put
- patch

如果需要使用一个路由来处理所有的请求方法，则可以调用 app.all()：

```
app.all('/', (req, resp) => {
  resp.send('请求首页');
});
```

4.3.2 路由路径

路由路径结合请求方法，定义了可以发出请求的地址。路由路径可以是字符串、字符串模式或正则表达式。

查询字符串（一般称为 GET 参数）不是路由路径的一部分。

（1）以下是一些基于字符串的路由路径的示例。

以下路由路径会将请求匹配到根路由/：

```
app.get('/', (req, resp) => {
  resp.send('主页');
});
```

以下路由路径会将请求匹配到/about：

```
app.get('/about', (req, resp) => {
  resp.send('关于页面');
});
```

以下路由路径可以将请求匹配到/random.txt 文件：

```
app.get('/random.txt', (req, resp) => {
  resp.send('random.txt');
});
```

字符串模式可以"认为"是正则表达式的子集，支持部分正则表达式语法。

（2）以下是一些基于字符串模式的路由路径示例。

以下路由路径将与 /acd 和 /abcd 相匹配。"?"号在正则表达式中代表"至多一个"，示例代码中就是匹配"至多一个 b"：

```
app.get('/ab?cd', (req, resp) => {
  resp.send('ab?cd');
});
```

以下路由路径将与 /abcd、/abbcd、/abbbcd 等等相匹配。+在正则表达式中代表"至少一个"，示例代码中就是匹配"至少一个 b"：

```
app.get('/ab+cd', (req, resp) => {
  resp.send('ab+cd');
});
```

以下路由路径将与 /abcd 和 /ad 相匹配。()在正则表达式中代表分组，示例代码中就是"要么有一个 bc，要么没有 bc"：

```
app.get('/a(bc)?d', (req, resp) => {
  resp.send('a(bc)?d');
});
```

（3）以下是一些基于正则表达式的路由路径示例。

以下路由路径将与任何带有 user 的请求链接相匹配：

```
app.get(/user/, (req, resp) => {
  resp.send('/user/');
});
```

以下路由路径将严格与 /admin 匹配：

```
app.get(/^\/admin$/, (req, resp) => {
```

```
resp.send('/^\/admin$/');
});
```

4.3.3 路由参数

路由参数用于捕获 URL 中各位置的值。捕获的值将填充到 req.params 对象中，并将路由路径中指定的 route 参数名称作为 req.params 对象的键。

```
app.get('/users/:userId/timelines/:timelineId', (req, resp) => {
  resp.json(req.params);
});
```

访问：http://localhost:8080/users/1/timelines/1，将得到以下响应：

```
{
  "userId": "1",
  "timelineId": "1"
}
```

路由参数的名称必须由 [A-Za-z0-9_]（也就是大小写字母、数字、下画线）组成。

字符串和字符串模式路由路径中的中划线 "-" 和点 "."无特殊意义，Express 按照字面意思处理这两个字符。因为在正则表达式中这两个字符有特殊意义，特此说明以防止混淆。

```
app.get('/users/:firstName.:lastName', (req, resp) => {
  resp.json(req.params);
});
```

访问 http://localhost:8080/users/lei.xia，将得到以下响应：

```
{
  "firstName": "lei",
  "lastName": "xia"
}
```

在 /users/:userId/timelines/:timelineId 例子中，预期匹配的是数字 ID 类型的参数，但是由于没有类型限制，字符串形式的参数也会匹配到。如果不限制参数类型，容易引发类型问题。

在路由参数后使用正则表达式可以限制参数类型，不满足类型的参数无法匹配该路由。

比如上述例子中，我们限制 userId 和 timelineId 为数字类型，可以使用以下代码：

```
app.get('/users/:userId(\\d+)/timelines/:timelineId(\\d+)', (req, resp) => {
    resp.json(req.params);
});
```

访问 http://localhost:8080/users/1/timelines/1，将得到以下响应：

```
{
  "userId": "1",
  "timelineId": "1"
}
```

访问 http://localhost:8080/users/1a/timelines/1，将得到以下响应：

```
Cannot GET /users/1a/timelines/1
```

因为 1a 与 (\d+) 不匹配。

4.3.4 路由函数

路由方法和路由路径匹配之后就会执行对应的路由函数，路由函数的签名如下：

```
function(request, response, next)
```

- request Express：请求对象。
- response Express：响应对象。
- next：匹配的下一个路由函数（可选参数）。

1. 单个路由函数

最简单的路由函数，对请求处理后发出响应，结束本次请求处理。

```
app.get('/', (req, resp) => {
  resp.send('/');
});
```

2. 多个路由函数

对于同一个路由，可以定义多个路由函数来处理，每个路由函数做一项工作。

不要忘记 next()方法的调用，即使定义多个路由函数，只要第一个函数未调用 next()，后续的路由函数都不会执行。

如下示例定义了两个路由函数，第一个函数打印了当前请求方法和请求路径，然后调用 next()执行下一个路由函数以响应此次请求。

```
app.get('/', (req, resp, next) => {
    console.log(`${req.method} ${req.path}`);
    next();
}, (req, resp) => {
    resp.send('首页');
});
```

在上面的示例中，访问首页时终端会输出"GET /"，之后会向客户端输出"首页"字样。

Express 中还可以使用函数数组来定义多个路由函数，上述例子的另外一种写法如下，两种写法效果是一样的。

```
function logger(req, resp, next) {
    console.log(`${req.method} ${req.path}`);
    next();
}

function home(req, resp) {
    resp.send('首页');
}
```

```
// 设置路由
app.get('/', [logger, home]);
```

3. 公共路由路径

如果多个路由有同样的路由路径，只是请求方法不同，若分别为每种方法定义一次路由，则不利于模块化。

```
app.get('/user/login', (req, resp) => {
  resp.send('登录页面');
});

app.post('/user/login', (req, resp) => {
  resp.send('登录处理');
});
```

针对这种典型场景，Express 提供了 app.router() 来处理：

```
app.route('/user/login')
    .get((req, resp) => {
       resp.send('登录页面');
    })
    .post((req, resp) => {
       resp.send('登录处理');
    });
```

4. 模块化的路由

使用 Express 提供的 Router 对象可以创建模块化的对象，实现路由和入口 JS 的解耦。使用模块化的路由有以下优点：

- 便于维护。
- 统一的路由前缀。
- 模块化。

user.js 用户相关路由：

```
const express = require('express');
const router = express.Router();

router.get('/login', (req, resp) => {
   resp.send('登录');
});

router.get('/register', (req, resp) => {
   resp.send('注册');
});

// 导出路由对象
module.exports = router;
```

timeline.js 动态相关路由：

```
const express = require('express');
const router = express.Router();

router.get('/list', (req, resp) => {
    resp.send('动态列表');
});

// 导出路由对象
module.exports = router;
```

index.js 入口文件:

```
const express = require('express');
const app = express();

const user = require('./user');
const timeline = require('./timeline');

app.use('/user', user);              // 设置路由前缀
app.use('/timeline', timeline);      // 设置路由前缀

// 开启监听
app.listen(8080, () => {
    console.log('listen on 8080');
});
```

上述例子最终会生成以下路由:

- GET /user/login。
- GET /user/register。
- GET /timeline/list。

4.4 请求对象

每个路由函数都会接收一个 request 对象,通过该对象可以获取本次请求的一些信息,比如请求方法、请求路径、请求参数等等。

表 4-1 是 request 对象常用的属性说明。通过"request.属性名"可以访问对应的属性。

表 4-1 request 对象常用的属性

属性名	说明	备注
method	请求方法	
path	请求路径	GET 参数不包含在 path 中
url	请求 URL	除域名外的完整 URL,包含 GET 参数

（续表）

属性名	说明	备注
query	GET 参数对象	
params	路由参数对象	
headers	请求报头	
cookies	请求 cookie	需要使用 cookie-parser 中间件
ip	客户端 IP	
body	POST 请求数据	需要使用 body-parser 中间件

```
app.get('/user/:userId', (req, resp) => {
    resp.json({
        method: req.method,
        path: req.path,
        url: req.url,
        query: req.query,
        params: req.params,
        headers: req.headers,
        cookies: req.cookies,
        ip: req.ip || req.ips,
    });
});
```

访问 http://localhost:8080/user/1?name=xialei&test=1，将响应以下 JSON：

```
{
  "method": "GET",
  "path": "/user/1",
  "url": "/user/1?name=xialei&test=1",
  "query": {
    "name": "xialei",
    "test": "1"
  },
  "params": {
    "userId": "1"
  },
  "headers": {
    "host": "localhost:8080",
    "connection": "keep-alive",
    "dnt": "1",
    "upgrade-insecure-requests": "1",
    "user-agent": "Mozilla/5.0 (Macintosh; Intel Mac OS X 10_14_5) AppleWebKit/537.36 (KHTML, like Gecko) Chrome/78.0.3904.87 Safari/537.36",
    "sec-fetch-user": "?1",
    "accept":
    "text/html,application/xhtml+xml,application/xml;q=0.9,image/webp,image/apng,*/*;q=0.8,application/signed-exchange;v=b3",
```

```
    "sec-fetch-site": "none",
    "sec-fetch-mode": "navigate",
    "accept-encoding": "gzip, deflate, br",
    "accept-language": "zh-CN,zh;q=0.9,en;q=0.8,zh-TW;q=0.7",
    "cookie": "Hm_lvt_78141f99bbb49f1564a7af89344f5acd=1500000000"
  },
  "ip": "::1"
}
```

4.4.1 获取请求 Cookie

Express 默认不解析请求报头中的 Cookie。如果需要获取 Cookie，可以自己开发中间件或者使用 cookie-parser 中间件。

中间件是可以访问请求对象、响应对象以及 next()的函数。中间件可以完成以下任务：

- 执行任何代码。
- 更改请求和响应对象。
- 结束请求处理。
- 调用下一个中间件。

比如 cookie-parser 中间件本质上还是解析 header 中 cookie 请求报头，将其挂载到 request.cookies 上。关于中间件的内容将在后面的内容详细介绍，这里我们先直接使用。

安装 cookie-parser 中间件：

```
npm install cookie-parser --save
// 导入 express 模块
const express = require('express');
const cookieParser = require('cookie-parser');
// 创建应用
const app = express();
app.use(cookieParser()); // 使用中间件

app.get('/', (req, resp) => {
    resp.json({ cookies: req.cookies });
});
// 开启监听
app.listen(8080, () => {
    console.log('listen on 8080');
});
```

访问 http://localhost:8080/，将响应以下 JSON：

```
{
  "cookies": {
    "Hm_lvt_78141f99bbb49f1564a7af89344f5acd": "1500000000"
  }
}
```

4.4.2 获取请求体

Express 默认也不处理请求体，如果需要获取请求体，需要监听 request 的 data 和 end 事件手动解析，这里我们直接使用 body-parser 中间件即可。

安装 body-parser 中间件：

```
npm install body-parser -save
// 导入 express 模块
const express = require('express');
const bodyParser = require('body-parser');
const app = express();

app.use(bodyParser()); // 使用中间件

app.post('/', (req, resp) => {
   resp.json(req.body);
});
// 开启监听
app.listen(8080, () => {
   console.log('listen on 8080');
});
```

使用 PostMan 向 http://localhost:8080 发起 POST 请求，请求 JSON 如下：

```
{
   "name": "xialei"
}
```

服务端会响应如下 JSON：

```
{
   "name": "xialei"
}
```

PostMan 是一个网络工具，可以方便地进行接口请求与调试。下载地址：https://www.getpostman.com/downloads/。

4.5 响应对象

每个路由函数都会接收一个响应对象，通过调用响应对象的方法，可以将响应发送到客户端，并结束请求处理。如果路由函数未调用响应对象的任何方法，则客户端请求将被挂起，直到客户端超时。

下面的内容介绍了响应对象常用的操作。

1. resp.status()

设置响应状态码。

```
resp.status(statusCode);
```

- statusCode 响应状态码

以下是响应 403 状态码的示例：

```
app.get('/', (req, resp) => {
  resp.status(403).end();
});
```

2. resp.set()

设置响应报头。要一次设置多个响应报头字段，需要传递对象作为参数。

```
resp.set(field[, value])
```

或

```
resp.set({
  [field]: value
});
```

- field：响应报头字段名称。
- value：响应报头字段值。

以下是设置响应类型为纯文本的示例：

```
app.get('/', (req, resp) => {
  resp.set('Content-Type', 'text/plain');
  resp.send('Hello World');
});
```

3. resp.download()

通过 Content-Disposition 响应报头提示客户端下载文件。

```
resp.download(path[, filename][, options][, callback])
```

- path：需要提供给客户端下载的服务端文件路径。
- filename：客户端下载文件时的别名。
- options：下载选项。
- callback：回调函数。

以下是一个下载 PDF 的示例：

```
app.get('/download', (req, resp) => {
  resp.download('./data.pdf', 'data.pdf', (err) => {
    if(err) {
      console.warn('下载失败', err);
    }
  });
});
```

```
});
```

4. resp.end()

结束响应过程。

```
resp.end([chunk][, encoding][, callback])
```

- chunk：响应数据。
- encoding：响应体编码。
- callback：回调函数。

以下是一个输出 Hello World 并结束请求的示例：

```
app.get('/', (req, resp) => {
  resp.end('Hello World');
});
```

5. resp.redirect()

重定向到指定的 URL Path 或者完整的 URL 链接。默认情况下响应状态码为 302。

```
resp.redirect([status,] path)
```

- status：响应状态码，301 或者 302。
- path：重定向路径。

以下是重定向到登录页面的示例：

```
app.get('/user/home', (req, resp) => {
  resp.redirect('/user/login');
});
```

以下是使用 301（永久重定向）状态码将文章重定向的示例：

```
app.get('/news/2019/10/01.html', (req, resp) => {
  resp.redirect(301, '/news/2019/10-01.html');
});
```

以下是重定向到外部网域的示例：

```
app.get('/', (req, resp) => {
  resp.redirect('https://www.ddhigh.com');
});
```

6. resp.render()

渲染 HTML 模板页面。
渲染模板页面需要使用到模板引擎，关于模板的详细内容将在后续内容介绍。

```
resp.render(view[,locals][,callback])
```

- view：视图名称。
- locals：传递到视图的变量对象，视图可以访问到这些变量并进行渲染。
- callback：回调函数。如果提供，该方法将返回可能的错误和 HTML 字符串，但不是自动发送 HTML 到客户端。

以下是渲染用户主页的示例：

```
app.get('/user/home', (req, resp) => {
   resp.render('user/home', { name: 'xialei' });
});
```

7. resp.cookie()

设置 Cookie。

```
resp.cookie(name, value[, options])
```

- name： Cookie 名称。
- value： Cookie 值。
- options 选项：
    - domain：域名。默认为当前域名。
    - expires GMT：到期时间。如果未设置或者设置为 0，则浏览器关闭后 Cookie 失效。
    - httpOnly：将 Cookie 标记为只有 HTTP 服务器能访问（客户端 JS 无法访问）。
    - maxAge：以毫秒为单位的过期时间。通常比 expires 选项使用方便。
    - path：Cookie 路径。
    - secure：标记为仅在 HTTPS 协议下才发送。
    - signed：是否对 Cookie 签名。

以下是设置用户登录态 cookie 的示例：

```
app.get('/', (req, resp) => {
   resp
      .cookie('logged', 'true', {
         httpOnly: true,
         maxAge: 24 * 3600 * 1000      // 有效期1天
      })
      .end();
});
```

调用 cookie 并不会终止请求处理流程，需要调用其他方法来终止请求处理，否则客户端请求会挂起直到超时。

8. resp.send()

发送 HTTP 响应。该方法可以根据传入的内容来输出不同格式的响应内容。

```
resp.send([body])
```

- body 响应内容。支持 string、Buffer、Object、Array。
    - String：Content-Type 将自动设置为 text/html。
    - Buffer：Content-Type 将自动设置为 application/octet-stream。
    - Object 或 Array：Content-Type 将自动设置为 application/json。

以下是响应 JSON 的示例：

```
app.get('/', (req, resp) => {
  resp.send({name: 'xialei'});
});
```

以下是响应文本的示例：

```
app.get('/', (req, resp) => {
  resp.send('Hello World');
});
```

4.6 中间件

中间件是一个函数，在应用的请求-响应周期中，能够访问请求对象、响应对象和 next 函数。中间件可以执行以下任务：

- 执行逻辑代码。
- 更改请求和响应对象。
- 结束请求-响应周期。
- 调用下一个中间件。

如果当前的中间件没有调用 next()，也没有结束请求-响应周期，则该请求将被挂起。

中间件有路由中间件和全局中间件两种。全局中间件对任何请求都会生效，路由中间件只对特定的路由生效。

4.6.1 全局中间件

以下是全局的日志中间件示例，该中间件打印当前的请求方法、请求路径以及 User-Agent。

```
// 导入 express 模块
const express = require('express');
const app = express();

function logger(req, resp, next) {
    console.log(`${req.method} ${req.path} "${req.headers['user-agent']}"`);
    next();
}

app.use(logger);

app.get('/', (req, resp) => {
    resp.send('Hello World');
});
app.get('/user', (req, resp) => {
  resp.send('user');
})
```

```
// 开启监听
app.listen(8080, () => {
    console.log('listen on 8080');
});
```

访问 http://localhost:8080，客户端会收到 Hello World 响应，并且在终端会输出以下日志：

GET / "Mozilla/5.0 (Macintosh; Intel Mac OS X 10_14_5) AppleWebKit/537.36 (KHTML, like Gecko) Chrome/78.0.3904.87 Safari/537.36"

GET /favicon.ico "Mozilla/5.0 (Macintosh; Intel Mac OS X 10_14_5) AppleWebKit/537.36 (KHTML, like Gecko) Chrome/78.0.3904.87 Safari/537.36"

访问 http://localhost:8080/user，客户端会收到 user 响应，终端同样会输出请求日志。

4.6.2 路由中间件

还是以刚才的中间件为例，不过我们将其挂载到特定的路由上。

```
// 导入 express 模块
const express = require('express');
const app = express();

function logger(req, resp, next) {
    console.log(`${req.method} ${req.path} "${req.headers['user-agent']}"`);
    next();
}

app.get('/', logger, (req, resp) => {    // 挂载到首页
    resp.send('Hello World');
});

app.get('/user', (req, resp) => {        // 未使用中间件
    resp.send('user');
});

// 开启监听
app.listen(8080, () => {
    console.log('listen on 8080');
});
```

访问 http://localhost:8080，终端会输出请求日志，但是访问 http://localhost:8080/user 时终端不会输出请求日志。

4.6.3 可配置的中间件

在请求对象这一节中，我们使用了如下代码来挂载中间件。

```
app.use(cookieParser());
```

在上面的内容中，我们使用的下面代码。

```
app.use(logger);
```

cookieParser()是函数调用，支持传入配置，并且返回一个中间件。cookieParser()的代码结构如下：

```
function cookieParser(options) {
  return function(req, resp, next) {

  };
}
```

我们针对日志中间件做一下修改来支持配置：

```
// 导入express 模块
const express = require('express');
const app = express();

function logger(options) { // 配置选项
    return function (req, resp, next) { // 返回真正的中间件
        const logs = [];
        if (options.method) {
            logs.push(req.method);
        }
        if (options.path) {
            logs.push(req.path);
        }
        if (options['user-agent']) {
            logs.push(req.headers['user-agent']);
        }
        console.log(logs.join(' '));
        next();
    };
}

app.use(logger({ method: true, path: truc })); // 使用中间件，并传入配置

app.get('/', (req, resp) => {
  resp.send('Hello World');
});

app.get('/user', (req, resp) => {
   resp.send('user');
});

// 开启监听
app.listen(8080, () => {
   console.log('listen on 8080');
});
```

4.6.4 Cookie 中间件

在前面的内容中我们使用 cookie-parser 来读取 cookie，本节将和大家一起来编写一个 cookie-parser。

通过请求报头可以获取原始的 cookie 请求报头：

```
{
  "cookie": "Idea-26226ea6=760672c4-a540-4189-a6d4-8d71c8b48f;
    grafana_session=20e77e60cad3c900ad12cf8d6dc2c1;
    Hm_lvt_78141f99bbb49f1564a7af89344f5a=1570785976,1572421010,1573309342"
}
```

可以看到每个 cookie 之间以分号分隔，cookie 的名字和 cookie 值通过等号分隔，所以编写 cookie 的思路如下：

（1）读取请求报头的 cookie 字段。
（2）使用";"来分隔每个 cookie。
（3）针对单个 cookie，使用"="分隔名字和值。
（4）将获取到的 cookie 名字和值挂载到 req.cookies 对象上。

```
function cookieParser() {
    return function (req, resp, next) {
        req.cookies = {}; // 定义 cookie 对象
        const headerCookie = req.headers.cookie;    // 获取请求报头的 cookie 字段
        if (headerCookie) {
            const cookies = headerCookie.split(';');    // 分隔每个 cookie
            cookies.forEach((cookie) => {
                const pairs = cookie.split('=');    // 分隔单个 cookie 的名字和值
                // 将解析到的 cookie 挂载到 req.cookies 对象上
                req.cookies[pairs[0]] = pairs[1];
            });
        }
        next();
    }
}

app.use(cookieParser());          // 使用中间件

app.get('/', (req, resp) => {
    resp.send(req.cookies);       // 输出 cookie
});
```

4.6.5 响应时长中间件

有些场景下需要统计请求开始到请求结束阶段，服务器处理请求所花费的时长，目前的中间件是无法获取到"请求结束"事件的。

Express 的响应对象提供了"finish"事件来告知请求结束事件。我们基于该事件来编写响应时长中间件。

```
function responseTime(req, resp, next) {
    const start = Date.now();                    // 开始处理时间
    resp.once('finish', function () {            // 处理结束时间
        console.log('处理时长', Date.now() - start);
    });
    next();
}
app.use(responseTime);  // 挂载中间件

app.get('/', (req, resp) => {
    setTimeout(() => {
        resp.send(req.cookies);
    }, 1000);
});
```

访问 http://localhost:8080 时，终端会打印响应时长。

监听 resp 事件需要使用 once，使用 on 会引起内存泄露。

4.6.6 静态资源中间件

为了提供图片、CSS 和 JS 之类的静态文件的访问，可以使用内置的 express.static 中间件。

```
express.static(root, [options])
```

- root：服务器文件夹路径。
- options：选项。

以下是一个提供静态资源访问的示例。

项目目录结构如下：

|---index.js

|---public

　　|---css

　　　　|---style.css

```
// 导入 express 模块
const express = require('express');
const app = express();

app.use(express.static('./public'));  // 将静态资源中间件挂载到全局

// 开启监听
app.listen(8080, () => {
```

```
    console.log('listen on 8080');
});
```

上面的代码需要访问 http://localhost:8080/css/style.css 才能获取 style.css。

也就是说 public 目录在请求链接中不能出现，如果需要将静态资源挂载到 /public 这个路由前缀下，可以使用下面的代码：

```
app.use('/public', express.static('./public'));
```

上面的代码需要访问 http://localhost:8080/public/css/style.css 才能获取 style.css。

4.7　错误处理

错误处理指 Express 如何捕获和处理同步和异步发生的错误。Express 带有默认错误处理程序，因此一般情况下无须手动编写错误处理程序。

4.7.1　同步错误

以下是一个同步错误的示例：

```
app.get('/', (req, resp) => {
  if(!req.query.name) {
    throw new Error('缺少参数name');
  }
  resp.end('ok');
});
```

访问 http://localhost:8080，将提示如下错误信息，这是 Express 默认的模板。

```
Error: 缺少参数 name
    at /Users/xialeistudio/WebstormProjects/nodejs-inaction-examples/
    express-example/index.js:8:15
    at Layer.handle [as handle_request]
    (/Users/xialeistudio/WebstormProjects/nodejs-inaction-examples/express-exa
    mple/node_modules/express/lib/router/layer.js:95:5)
    at next (/Users/xialeistudio/WebstormProjects/nodejs-inaction-examples/
    express-example/node_modules/express/lib/router/route.js:137:13)
    at Route.dispatch (/Users/xialeistudio/WebstormProjects/nodejs-inaction-
    examples/express-example/node_modules/express/lib/router/route.js:112:3)
    at Layer.handle [as handle_request]
    (/Users/xialeistudio/WebstormProjects/nodejs-inaction-examples/express-exa
    mple/node_modules/express/lib/router/layer.js:95:5)
    at /Users/xialeistudio/WebstormProjects/nodejs-inaction-examples/
    express-example/node_modules/express/lib/router/index.js:281:22
    at Function.process_params (/Users/xialeistudio/WebstormProjects/
    nodejs-inaction-examples/express-example/node_modules/express/lib/router/i
```

```
ndex.js:335:12)
    at next (/Users/xialeistudio/WebstormProjects/nodejs-inaction-examples/
express-example/node_modules/express/lib/router/index.js:275:10)
    at expressInit (/Users/xialeistudio/WebstormProjects/nodejs-
inaction-examples/express-example/node_modules/express/lib/middleware/init
.js:40:5)
    at Layer.handle [as handle_request] (/Users/xialeistudio/WebstormProjects/
nodejs-inaction-examples/express-example/node_modules/express/lib/router/l
ayer.js:95:5)
```

同步错误一般交给框架自动处理即可。

4.7.2 异步错误

异步错误一般是发生在回调函数中的错误，需要通过 next(err) 才能捕获异步错误。

以下是捕获读取文件的错误示例：

```
app.get('/', (req, resp, next) => {
    fs.readFile('./a.js', { encoding: 'utf8' }, (err, data) => {
        if (err) {
            next(err);
            return;
        }
        resp.end(data);
    });
});
```

异步错误需要手动调用 next() 并传递 Error 对象。

传递给 next() 的参数可以是字符串、数字、Error 对象等等，建议传递 Error 对象。只要传递了非空参数，Express 就会执行错误处理流程。

4.7.3 自定义错误处理函数

如果需要自定义错误函数的逻辑，可以使用自己编写的错误处理函数。

错误处理函数的签名如下：

```
function errorHandler(err, req, resp, next)
```

- err：错误对象。
- req：请求对象。
- resp：响应对象。
- next：下一个错误处理器。

处理器本质也是中间件，但是需要放置在所有中间件、路由函数的后面才会生效。

以下是将默认的 HTML 错误响应更改为 JSON 的示例。

```
// 导入express模块
```

```
const express = require('express');
const fs = require('fs');
const app = express();


app.get('/', (req, resp, next) => { // 路由函数
    throw new Error('发生错误');
});

app.use((err, req, resp, next) => { // 错误处理
    resp.json({
        path: req.path,
        message: err.message
    });
});

// 开启监听
app.listen(8080, () => {
    console.log('listen on 8080');
});
```

4.7.4 多个错误处理函数

函数式编程一个重要的哲学是：一个函数只做一件事。以错误处理来说，我们需要记录日志、发送响应，这两件事可以通过两个中间件来完成：一个记录日志；另一个发送响应。

```
app.get('/', (req, resp, next) => { // 路由函数
    throw new Error('发生错误');
});
// 记录日志
app.use((err, req, resp, next) => {
    fs.writeFile('./app.log', `${req.method} ${req.url} Error: ${err.message}`,
    (err) => {
        next(err); // 交给下一个错误处理函数处理
    });
});
// 响应 JSON
app.use((err, req, resp, next) => {
    resp.json({
        path: req.path,
        message: err.message
    });
});
```

多个错误处理函数工作时不要忘记调用 next()，否则不会进入下一个错误处理函数导致请求挂起。

4.8 模板渲染

模板引擎能够在应用中使用静态模板文件，在运行时，可以根据实际变量替换模板中的变量，并将渲染后的 HTML 发送给客户端。

Express 可以使用 jade、pug、mustache 和 ejs 作为模板引擎，默认的是 jade，本书中使用 ejs 作为模板引擎。

要渲染模板文件，需要安装对应的模板引擎，还需要更改 app 设置。

```
app.set('views', './templates');      // 设置模板文件目录
app.set('view engine', 'ejs');        // 设置模板引擎
```

4.8.1 使用 ejs 模板

ejs 是接近普通 HTML 语言的模板，通过标记来赋予 HTML 模板动态化的能力。

使用 npm 安装 ejs 模块：

```
npm install ejs -save
```

设置应用模板配置：

```
app.set('views', './templates');
app.set('view engine', 'ejs');
```

以下是完整的示例：

```
// index.js
const express = require('express');
const app = express();

app.set('views', './templates');
app.set('view engine', 'ejs');

app.get('/', (req, resp) => {
   resp.render('site/index', {
       title: '首页'
   });
});

// 开启监听
app.listen(8080, () => {
   console.log('listen on 8080');
});
```

渲染模板时传递了 title 变量，可以在模板中获取到：

```
// templates/site/index.ejs
```

```
<!DOCTYPE html>
<html lang="en">
<head>
    <meta charset="UTF-8">
    <meta name="viewport" content="width=device-width, initial-scale=1.0">
    <meta http-equiv="X-UA-Compatible" content="ie=edge">
    <title><%= title %></title>
</head>
<body>

</body>
</html>
```

<%= title%> 是渲染变量的语法，title 是在路由函数中传递的。

4.8.2 ejs 语法

ejs 的语法比较多，这里列举比较常用的语法。

转义输出，如果 title 字符串含有 HTML 代码，则最终显示时会被转义为实体字符。

```
<%= title %>
```

不转义输出，不转义 HTML 代码，存在 XSS（一种客户端 JS 攻击手段）风险。

```
<%- title %>
```

（1）执行 JS 代码：

```
<% 代码 %>
```

（2）导入其他模板：

```
<% include 模板路径 %>
```

（3）逻辑判断：

```
<% if(condition) { %>
// HTML 代码
<% } %>
```

（4）循环：

```
<% list.forEach((item) => { %>
  <%= item %>
<% }) %>
```

以下是一个完整的示例：

```
// inde.js
const express = require('express');
const app = express();

app.set('views', './templates');
```

```js
app.set('view engine', 'ejs');

app.get('/', (req, resp) => {
    const users = [
        { id: 1, name: '张三' },
        { id: 2, name: '夏磊' }
    ];
    resp.render('site/index', {
        users: users,
        show: !!req.query.show
    });
});

// 开启监听
app.listen(8080, () => {
    console.log('listen on 8080');
});
```

```ejs
<!-- templates/site/index.ejs -->
<!DOCTYPE html>
<html lang="en">

<head>
    <meta charset="UTF-8">
    <meta name="viewport" content="width=device-width, initial-scale=1.0">
    <meta http-equiv="X-UA-Compatible" content="ie=edge">
    <title>用户列表</title>
</head>

<body>
    <% if(show) { %> <!-- 判断 -->
    <ul>
        <% users.forEach((user) => { %> <!-- 循环 -->
        <li><%= user.id%> - <%= user.name%></li>
        <% }) %>
    </ul>
    <% } %>
</body>

</html>
```

访问 http://localhost:8080 时，页面标题为"用户列表"，页面主体为空白。

访问 http://localhost:8080?show=1 时，页面标题为"用户列表"，页面主体如下：

1 - 张三

2 - 夏磊

4.9 留言板项目开发

本节将和大家一起使用所学的知识开发一个 Web 留言板。

项目的功能如下：

- 首页展示留言列表和发布按钮。
- 发布页面展示留言表单，单击发布后保存留言。

本节的留言数据存储在内存中，进程结束留言板数据会丢失。数据库相关的知识在后面的章节中介绍。

4.9.1 开始编码

初始化项目并安装相关依赖：

```
mkdir express-board
cd express-board
npm init -y        # 直接使用默认选项回答所有问题
npm install body-parser express ejs -save
```

下面列出本项目的完整代码。

index.js

```
const express = require('express');
const bodyParser = require('body-parser');
const app = express();

// 模板引擎配置
app.set('view engine', 'ejs');
app.set('views', './templates');
// 请求体解析中间件配置
app.use(bodyParser());

// 留言内容
const messages = [];

// 首页路由
app.get('/', (req, resp) => {
    resp.render('index', { messages });
});

// 留言页路由
app.route('/publish')
    .get((req, resp) => {
        resp.render('publish');
```

```js
    })
    .post((req, resp) => {
        if (!req.body.name || !req.body.content) {
            throw new Error('请将所有选项填写完整');
        }
        const now = (new Date()).toLocaleString();
        messages.push({
            name: req.body.name,
            content: req.body.content,
            time: now // 留言事件
        });

        resp.redirect('/');
    });

// 开启监听
app.listen(8080, () => {
    console.log('listen on 8080');
});
```

在 index.js 的同级新建 templates/index.ejs 来存放首页模板。

templates/index.ejs

```html
<!DOCTYPE html>
<html lang="en">

<head>
    <meta charset="UTF-8">
    <meta name="viewport" content="width=device-width, initial-scale=1.0">
    <meta http-equiv="X-UA-Compatible" content="ie=edge">
    <title>留言列表</title>
</head>

<body>
    <a href="/publish">发表留言</a>

    <% if(messages.length === 0) { %>
    <p>当前没有留言。</p>
    <% } else { %>
    <table>
        <tr>
            <th>留言时间</th>
            <th>留言人</th>
            <th>留言内容</th>
        </tr>
        <% messages.forEach((message) => { %>
        <tr>
            <td><%= message.time%></td>
            <td><%= message.name%></td>
            <td><%= message.content%></td>
        </tr>
```

```
        <% }) %>
    </table>
<% } %>
</body>
</html>
```

该模板文件对留言内容进行了判空处理，没有留言内容就显示"当前没有留言。"，有留言内容就显示留言列表。

在 index.js 的同级新建 templates/publish.ejs 来存放发布页模板。

```
templates/publish.ejs
<!DOCTYPE html>
<html lang="en">
<head>
    <meta charset="UTF-8">
    <meta name="viewport" content="width=device-width, initial-scale=1.0">
    <meta http-equiv="X-UA-Compatible" content="ie=edge">
    <title>发表留言</title>
</head>
<body>
    <form action="/publish" method="POST"
    enctype="application/x-www-form-urlencoded">
        <fieldset>
            <legend>发表留言</legend>
            <div>
                <label for="name">姓名</label>
                <input type="text" name="name" id="name" required>
            </div>
            <div>
                <label for="content">内容</label>
                <textarea name="content"></textarea>
            </div>
            <div>
                <button type="submit">发表</button>
                <button type="reset">重置</button>
            </div>
        </fieldset>
    </form>
</body>
</html>
```

4.9.2 运行项目

node inde.js

浏览器访问 http://localhost:8080，首页效果如图 4-1 所示。

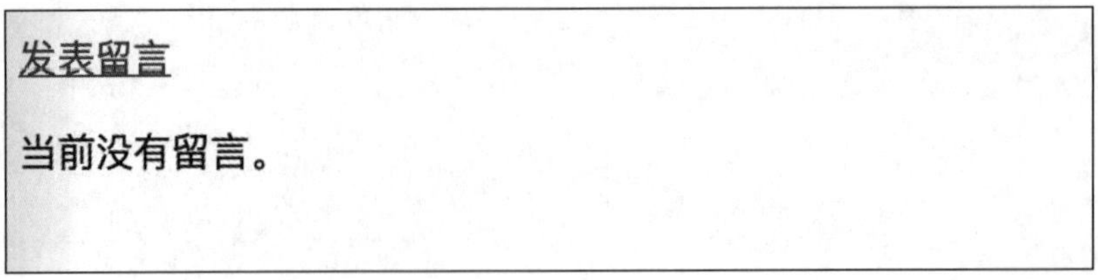

图 4-1

单击【发表留言】,发布页效果如图 4-2 所示。

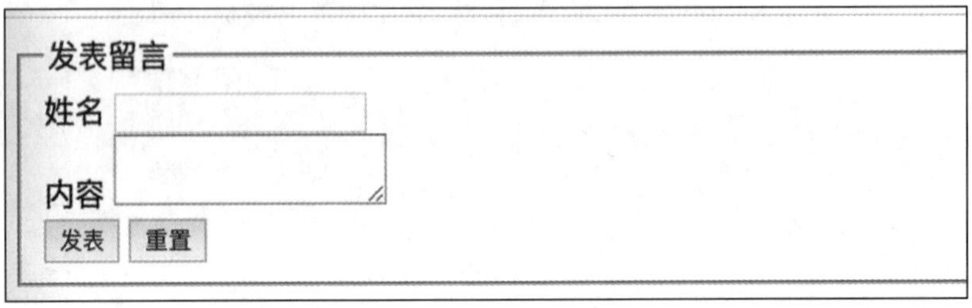

图 4-2

输入姓名和内容后,单击【发表】自动重定向到首页,首页效果如图 4-3 所示。

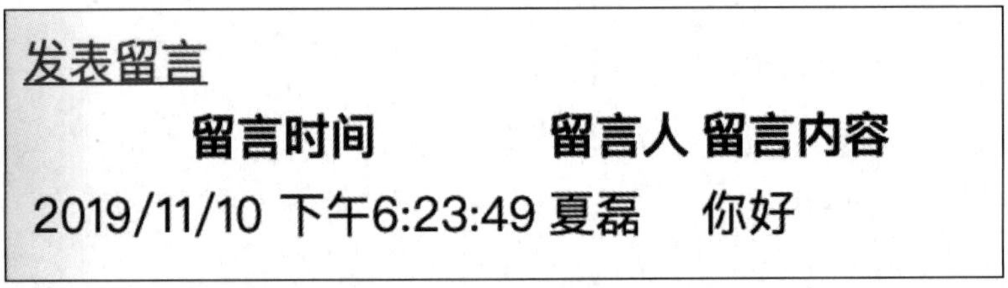

图 4-3

恭喜!你的第一个 Express 应用已经运行成功了!如果遇到问题可以咨询作者的公众号。
回顾一下一个 Express 应用的开发流程:

(1)初始化项目。
(2)安装依赖。
(3)编写路由文件,并导入中间件。
(4)编写模板文件。
(5)运行项目并测试。

4.10　本章小结

本章讲解了 Express 框架以及 Express 的核心技术,最后介绍了一个完整的 Express 应用开发。

回顾一下本章所学:

- Express 的项目结构。
- Express 的路由。
- Express 的请求与响应对象。
- Express 的中间件和错误处理。
- Express 处理静态资源。
- Express 的模板渲染。
- Express 的项目实战。

下一章我们将学习以中间件开发为核心思想的 Koa 框架，这也是 Express 团队打造的下一代 Web 开发框架。

第 5 章

下一代 Web 开发框架——Koa

本章内容

- Koa 的基础使用
- Koa 的上下文
- Koa 的中间件
- Koa 的页面渲染与数据输出
- 基于 Koa 开发博客系统

5.1 Koa 简介

Koa 是由 Express 开发团队打造的一种全新的 Web 框架，旨在为 Web 应用提供更小、更强大的基础。通过基于 Promise 的异步编程，Koa 应用可以不使用回调（callback），大大提高了开发效率。此外，Koa 在其核心并未捆绑任何中间件（甚至于路由功能都需要外部中间件完成）。

由于 Koa 核心不捆绑任何中间件，因此 Koa 核心是"纯净的"，这极大地方便了用户扩展。此外，Koa 使用了 Promise、async/await 语法来进行异步编程，而 Express 是基于事件和回调的。

Koa 框架和 Express 框架的主要差别在于异步编程和中间件方面，其他特性是相似的。

- Express 框架使用回调来进行异步处理，这也是 Node.js 标准的做法。但是基于回调的异步编程模型在多个异步操作之间有顺序依赖时，会产生回调地狱（Callback Hell），也就是多层 callback 嵌套问题，代码不利于维护。

- Koa 框架使用了 ES2017 最新的 async/await 语法来进行异步编程，从根本上解决了传统 Node.js 异步编程风格存在的问题，但是需要将异步调用包装为 Promise，之后的内容我们将使用社区最强大的 Bluebird 来解决 Promise 包装问题。

由于 Koa 进行异步调用时强制使用 async/await，因此需要将异步回调方法转换为 Promise，如果每个回调方法都需要自己包装的话，工作量还是有点大，因此接下来将介绍这一问题目前最好的解决方案——Bluebird。

5.2 Bluebird

Bluebird 是 Node.js 最出名的 Promise 实现，除了实现标准的 Promise 规范之外，Bluebird 还提供了包装方法，可以快速地将 Node.js 回调风格的函数包装为 Promise。

Node.js 回调风格的函数如下：

```
function(err, data1, data2, ..., dataN)
```

回调函数的第 1 个参数永远为 Error 对象，如果出现错误，则 err 值是 Error 对象；如果未出现错误，则 err 值为 null。

Bluebird 的使用

使用 Bluebird 模块前需要使用 npm 安装：

```
npm install bluebird --save
```

bluebird.promisifyAll()可以将 Node.js 回调风格的函数包装为 Promise 函数，该方法签名如下：

```
bluebird.promisifyAll(target, options)
```

- target 需要包装的对象。如果 target 是普通对象，则包装后生成的异步 API 只有该对象持有；如果 target 是原型对象，则包装后生成的异步 API 被所有实例持有。
- options 选项：
  - suffix：异步 API 方法名后缀，默认为"Async"。如 fs.readFile()函数包装后生成的异步 API 为 fs.readFileAsync。
  - multiArgs：是否允许多个回调参数，默认 false。我们知道 Promise 的 then()方法只接受一个参数，而 callback 则可以回调多个参数，multiArgs 为 true 时，bluebird 将所有 callback 参数传入一个数组，Promise.then()接受该数组，从而得到多个参数。

bluebird.promisifyAll()只会给目标对象添加新方法，原来的 Node.js 回调风格的方法不受影响。

包装之后的方法和包装之前的方法使用起来只有一个差别，那就是不要传递回调函数，通过 Promise 获取结果。

下面是包装 fs 对象的示例，fs 不是原型对象，也没有 fs 实例，因此直接包装 fs 对象即可。

```
const bluebird = require('bluebird');
const fs = require('fs');

bluebird.promisifyAll(fs);

// 回调函数示例
fs.readFile('./data.log', {encoding:'utf8'}, (err, data) => {
  if(err) {
    console.warn(err);
    return;
  }
  console.log(data);
});

// Promise 示例
fs.readFileAsync('./data.log', {encoding:'utf8'}).then((data) => {
  console.log(data);
}).catch((err) => {
  console.warn(err);
});
```

以上两种方法的结果没有区别，通过 Promise 包装，可以批量地将对象进行 Promise 处理，结合 async/await，可以极大地提升异步编程体验。

5.3　Koa 快速开始

本节将和大家一起从零开始构建一个 Koa 应用，理清 Koa 应用的开发流程。

5.3.1　初始化项目

新建应用目录，然后进入该目录并将其作为工作目录：

```
mkdir koa-example
cd koa-example
```

通过 npm 创建一个 package.json 文件：

```
npm init
```

此命令将要求你输出几个参数，例如应用的名称和版本。在本节的内容中，你可以直接按回车键接受默认值。

接下来安装 Koa 并将其保存到 package.json 的依赖列表中。

```
npm install koa -save
```

5.3.2 开始编码

```
// 导入模块
const Koa = require('koa');

// 实例化应用
const app = new Koa();

// 中间件
app.use(async (ctx) => {
    ctx.body = 'Hello World';
});

// 监听
app.listen(10000, () => {
    console.log('listen on 10000');
});
```

访问 http://localhost:10000，浏览器会输出 Hello World。

目前，编写 Koa 应用的时候没有使用路由这一概念，而是使用了中间件概念。前面的内容中提到了，Koa 核心不捆绑任何中间件，因此路由功能是没有的，也就是说不能根据请求路径和请求方法来返回不同的响应。

另外，Koa 的中间件功能和 Express 中间件是类似的，也可以访问请求对象、响应对象以及 next 函数。所以上例中使用了一个中间件，不管接收什么请求都会回复 Hello World。

5.4 Context

Context 在 Koa 应用中又称为"上下文"，该对象包含了 Koa 请求对象、Koa 响应对象和应用实例，Context 可以理解一个容器，该容器挂载了本次请求的请求对象和响应对象等信息。

中间件通过操作 Context 对象来获取请求信息，处理之后返回响应。

Context 实例有以下常用的几个属性或方法（*号所标记是常用属性或方法）：

- ctx.request: Koa 的请求对象，一般不直接使用，通过别名引用来访问。
- ctx.response: Koa 的响应对象，一般不直接使用，通过别名引用来访问。
- *ctx.state: 自定义数据存储，比如中间件需要往请求中挂载变量就可以存放在 ctx.state 中，后续中间件可以读取。
- *ctx.throw(): 抛出 HTTP 异常。
- ctx.headers: 请求报头，ctx.request.headers 的别名。
- ctx.method: 请求方法，ctx.request.method 的别名。
- ctx.url: 请求链接，ctx.request.url 的别名。

- ctx.path：请求路径，ctx.request.path 的别名。
- *ctx.query：解析后的 GET 参数对象，ctx.request.query 的别名。
- ctx.host：当前域名，ctx.request.host 的别名。
- *ctx.ip：客户端 IP，ctx.request.ip 的别名。
- ctx.ips：反向代理环境下的客户端 IP 列表，ctx.request.ips 的别名。
- ctx.get()：读取请求报头，ctx.request.get 的别名。
- *ctx.body：响应内容，支持字符串、对象、Buffer，ctx.response.body 的别名。
- *ctx.status：响应状态码，ctx.response.status 的别名。
- ctx.type：响应体类型，ctx.response.type 的别名。
- *ctx.redirect()：重定向，ctx.response.redirect 的别名。
- *ctx.set()：设置响应报头，ctx.response.set 的别名。

以下是一个显示当前请求信息并添加自定义响应报头的示例。

```javascript
// 导入模块
const Koa = require('koa');

// 实例化应用
const app = new Koa();
app.proxy = true;
// 中间件
app.use(async (ctx) => {
    ctx.set('x-version', '1.0.0');
    ctx.body = {
        method: ctx.method,
        path: ctx.path,
        url: ctx.url,
        query: ctx.query,
        headers: ctx.headers,
        ip: ctx.ip
    };
});

// 监听
app.listen(10000, () => {
    console.log('listen on 10000');
});
```

访问 http://localhost:10000/?a=1&b=2，浏览器响应如下 JSON：

```json
{
  "method": "GET",
  "path": "/",
  "url": "/?a=1&b=2",
  "query": {
    "a": "1",
    "b": "2"
  },
```

```
"headers": {
  "host": "localhost:10000",
  "connection": "keep-alive",
  "pragma": "no-cache",
  "cache-control": "no-cache",
  "dnt": "1",
  "upgrade-insecure-requests": "1",
  "user-agent": "Mozilla/5.0 (Macintosh; Intel Mac OS X 10_14_5) 
  AppleWebKit/537.36 (KHTML, like Gecko) Chrome/78.0.3904.87 Safari/537.36",
  "sec-fetch-user": "?1",
  "accept": 
  "text/html,application/xhtml+xml,application/xml;q=0.9,image/webp,image/ap
  ng,*/*;q=0.8,application/signed-exchange;v=b3",
  "sec-fetch-site": "none",
  "sec-fetch-mode": "navigate",
  "accept-encoding": "gzip, deflate, br",
  "accept-language": "zh-CN,zh;q=0.9,en;q=0.8,zh-TW;q=0.7"
},
"ip": "::1"
}
```

5.5 Cookie 操作

Cookie 是 Web 应用维持少量数据的一种手段，通过 Cookie，服务端可以表示用户以及用户身份。本节我们将学习一下 Koa 如何操作 Cookie。

5.5.1 Cookie 签名

由于 Cookie 存放在浏览器端，存在篡改风险，因此 Web 应用一般会在存放数据的时候同时存放一个签名 cookie，以保证 Cookie 内容不被篡改。

Koa 中需要配置 Cookie 签名密钥才能使用 Cookie 功能，否则将报错。

```
const Koa = require('koa');
const app = new Koa();
app.keys = ['signedKey'];    // 推荐使用随机字符串
```

5.5.2 写入 Cookie

```
// 导入模块
const Koa = require('koa');
const app = new Koa();
app.keys = ['signedKey'];
app.use(async (ctx) => {
  ctx.cookies.set('logged', 1, {
```

```
        signed: true, // Cookie 签名
        httpOnly: true,
        maxAge: 3600 * 24 * 1000 // 有效期1天
    });
    ctx.body = 'ok';
});
// 监听
app.listen(10000, () => {
    console.log('listen on 10000');
});
```

访问 http://localhost:10000,查看浏览器的 Cookie 列表,结果如图 5-1 所示。

| logged | 1 | localhost | / | 2019-11… | 7 | √ |
| logged.sig | zyDik5pYKAdY4Ib4f1BTNCDfRKs | localhost | / | 2019-11… | 37 | √ |

图 5-1

可以看到除了我们设置的"logged"之外,多了个"logged.sig",这就是用来签名的 Cookie。服务端读取"logged"时,还会同时读取"logged.sig",一旦发现签名不匹配,则读取到的 cookie 值为"undefined"。

5.5.3 读取 Cookie

```
// 导入模块
const Koa = require('koa');
const app = new Koa();
app.keys = ['signedKey'];
app.use(async (ctx) => {
    const logged = ctx.cookies.get('logged', { signed: true });
    ctx.body = logged;
});
// 监听
app.listen(10000, () => {
    console.log('listen on 10000');
});
```

访问 http://localhost:10000,浏览器将显示"1"。

ctx.cookies.get()建议传递 signed 选项来验证签名,否则 cookie 将有篡改风险。

5.5.4 中间件

与 Express 类似,Koa 的中间件也能访问请求对象、响应对象和 next 函数,通常用来执行以下任务:

- 执行逻辑代码。
- 更改请求和响应对象。
- 结束请求-响应周期。

- 调用下一个中间件。
- 错误处理。

如果一个请求流程中，任何中间件都没有输出响应，Koa 中此次请求将返回 404 状态码（Express 会将请求挂起）。

造成这种差别的原因是 Express 需要手动执行输出函数才可以结束请求流程，而 Koa 使用了 async/await 来进行异步编程，不需要执行回调函数，直接对 ctx.body 赋值即可。

Koa 的中间件是一个标准的异步函数，函数签名如下：

```
async function middleware(ctx, next)
```

- ctx：上下文对象。
- next：下一个中间件。

运行完逻辑代码，将需要传递的数据挂载到 ctx.state，并且调用 await next() 才能将请求交给下一个中间件处理。

为了更好地理解中间件的执行流程，下面使用图 5-2 所示的模型来说明。

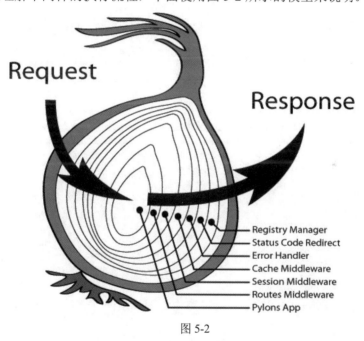

图 5-2

Koa 的中间件模型称为"洋葱圈模型"，请求从左边进入，有序地经过中间件处理，最终从右边输出响应。

最先 use 的中间件在最外层，最后 use 的中间件在最内层。

一般的中间件会执行两次（下面会给出示例），调用 next 之前为第一次，也就是"洋葱左半边"这一部分，从外层向内层依次执行。当后续没有中间件时，就进入响应流程，也就是"洋葱右半边"这一部分，从内层向外层依次执行，这是第二次执行。

下面举一个例子以方便大家理解该模型。

```js
// 导入模块
const Koa = require('koa');

// 实例化应用
const app = new Koa();

async function middleware1(ctx, next) {
    console.log('middleware1 start');
    await next();
    console.log('middlware1 end')
}
async function middleware2(ctx, next) {
    console.log('middleware2 start');
    await next();
    console.log('middlware2 end')
}
// 中间件
app.use(middleware1);
app.use(middleware2);
// 路由
app.use(async (ctx) => {
    console.log('router');
    ctx.body = 'Hello World';
});
// 监听
app.listen(10000, () => {
    console.log('listen on 10000');
});
```

访问 http://localhost:10000 时，终端输出如下：

```
middleware1 start
middleware2 start
router
middlware2 end
middlware1 end
```

也就是说，执行流程如下：

请求→中间件 1 开始→中间件 2 开始→路由处理→中间件 2 结束→中间件 1 结束→响应。

5.5.5 请求日志中间件

由于每个中间件有两次执行机会，因此相比于 Express 监听 finish 事件的方式，Koa 的中间件显得更加简单。

下面我们来开发一个日志中间件，该中间件会输出请求方法、请求路径、User-Agent 和处理时长。

```js
// 导入模块
const Koa = require('koa');
```

```
// 实例化应用
const app = new Koa();
// 日志中间件
async function logger(ctx, next) {
    const start = Date.now();
    await next();          // 将控制权交给下一个中间件
    console.log(`${ctx.method} ${ctx.path} "${ctx.headers['user-agent']}"
    ${Date.now() - start}ms`);
}
app.use(logger);
// "路由"
app.use(async (ctx) => {
    ctx.body = 'Hello World';
});
// 监听
app.listen(10000, () => {
    console.log('listen on 10000');
});
```

访问 http://localhost:10000，输出如下：

```
GET / "Mozilla/5.0 (Macintosh; Intel Mac OS X 10_14_5) AppleWebKit/537.36 (KHTML,
    like Gecko) Chrome/78.0.3904.87 Safari/537.36" 2ms
```

在 logger 中，我们在调用 next() 之前记录当前请求时间，调用 next() 后 logger 将不再执行，直接执行下一个中间件。本例中下一个中间件就是我们的路由了，路由执行完毕后，再来执行 logger 中间件 next() 之后的代码。

5.5.6 可配置的中间件

与 Express 类似，Koa 也可以通过"函数返回一个新函数"来编写可配置的中间件。

```
function middlewareName(options) {        // 选项
  return async function(ctx, next) {      // 真正的中间件

  }
}
```

下面通过配置选项来控制日志输出的内容：

```
// 导入模块
const Koa = require('koa');

// 实例化应用
const app = new Koa();
// 日志中间件
function logger(options) {                    // 选项
    return async function (ctx, next) {       // 真正的中间件
        const start = Date.now();
```

```js
        await next();
        const parts = [];
        options.method && parts.push(ctx.method);
        options.path && parts.push(ctx.path);
        options.userAgent && parts.push(ctx.headers['user-agent']);
        parts.push(`${Date.now() - start} ms`);
        console.log(parts.join(' '));
    }
}
app.use(logger({ method: true, path: true })); // 挂载中间件并传递选项
// "路由"
app.use(async (ctx) => {
    ctx.body = 'Hello World';
});
// 监听
app.listen(10000, () => {
    console.log('listen on 10000');
});
```

访问 http://localhost:10000，由于未开启 user-agent 来显示网页，因此终端输出如下：

```
GET / 2 ms
```

5.5.7 Cookie 解析中间件

与上面的中间件不同，本节编写的中间件解析完请求 Cookie 之后，需要将其挂载到 ctx.state 中，然后在路由中间件使用。

```js
// 导入模块
const Koa = require('koa');

// 实例化应用
const app = new Koa();
// 日志中间件
async function cookieParser(ctx, next) {
    const headerCookie = ctx.headers.cookie;
    ctx.state.cookies = {};
    if (headerCookie) {
        const cookies = headerCookie.split(';');
        cookies.forEach((cookie) => {
            const parts = cookie.split('=');
            ctx.state.cookies[parts[0]] = parts[1]; // 挂载到 ctx.state.cookies 下
        });
    }
    await next();
}
app.use(cookieParser);
// 路由
app.use(async (ctx) => {
    ctx.body = ctx.state.cookies;
```

```
});
// 监听
app.listen(10000, () => {
    console.log('listen on 10000');
});
```

访问 http://localhost:10000,输出如下:

```
{
  "Idea-26226ea6": "760672c4-a540-4189-a6d4-8d71c8b48fdd",
  "grafana_session": "20e77e60cad3c900ad12cf8d6dc2c1cb"
}
```

有些第三方中间件处理完数据并未挂载到 state 下,这不影响使用,但不是 Koa 推荐的做法,笔者建议各位读者还是将自定义的数据挂载到 ctx.state 下。

5.5.8 路由函数

Koa 核心并没有提供路由功能,但是可以使用一个默认的路由函数来提供响应。所有的请求都会执行该默认的路由函数。

路由函数的定义如下:

```
async function(ctx, next)
```

- ctx:请求上下文。
- next:下一个路由函数。

如果路由函数内部未使用异步逻辑,async 是可以省略的。

如下示例代码是合法且能执行的:

```
app.use((ctx) => {
  ctx.body = 'Hello World';
});
```

5.5.9 多个路由函数

一般来说,路由函数只有一个,设置响应数据到 ctx.body,执行完中间件后,请求终止。但是 Koa 支持同一个路由来使用多个路由函数。

```
// 导入模块
const Koa = require('koa');
const Router = require('koa-router');
// 实例化应用
const app = new Koa();
app.use(async (ctx, next) => { //路由函数1
    ctx.body = '1';
    await next();
});
```

```
app.use((ctx) => { // 路由函数 2
    ctx.body = '2';
});

// 监听
app.listen(10000, () => {
    console.log('listen on 10000');
});
```

访问 http://localhost:10000，将 "2" 输出到浏览器。

如果将路由函数 1 中的代码修改一下，将 "1" 输出到浏览器：

```
app.use(async (ctx, next) => {
    await next();
    ctx.body = '1';
});
```

因为 Koa 的洋葱圈模型，所以先执行路由函数 1，执行到 next() 后，进入路由函数 2 的执行，路由函数 2 设置了 ctx.body 为 "2"，此时没有后续路由函数，执行流程将回到路由函数 1，而此时路由函数 1 将 ctx.body 设置为 "1"，所以浏览器最终显示 "1"。

5.5.10 错误处理

Koa 有着简单且优雅的中间件机制，因此编写错误处理中间件变得很简单。

和 Express 错误处理中间件需要放置在应用末尾不同，Koa 采用了洋葱圈模型，所以 Koa 的错误处理中间件需要在应用的开始处挂载，这样才能将整个请求-响应周期涵盖，捕获其发生的错误。

```
// 导入模块
const Koa = require('koa');

// 实例化应用
const app = new Koa();
// 错误处理中间件
async function errorHandler(ctx, next) {
    try {
        await next();
    } catch (e) {
        ctx.status = e.status || 500;
        ctx.body = `System Error: ${e.message}`;
    }
}
app.use(errorHandler);

// 抛出错误
app.use(ctx => {
    ctx.throw(403, 'Forbidden');
});
```

```javascript
// 监听
app.listen(10000, () => {
   console.log('listen on 10000');
});
```

访问 http://localhost:1000 时，将输出"System Error: Forbidden"。

如果调换一下路由与错误处理中间件之间的位置呢？

```javascript
// 导入模块
const Koa = require('koa');

// 实例化应用
const app = new Koa();
// 错误处理中间件
async function errorHandler(ctx, next) {
   try {
      await next();
   } catch (e) {
      ctx.status = e.status || 500;
      ctx.body = `System Error: ${e.message}`;
   }
}

// 抛出错误
app.use(ctx => {
   ctx.throw(403, 'Forbidden');
});
app.use(errorHandler);

// 监听
app.listen(10000, () => {
   console.log('listen on 10000');
});
```

访问 http://localhost:10000，将输出"Forbidden"，也就是说错误处理器放在后面是不生效的，最终采用了 Koa 自带的错误处理机制。

5.5.11 多个错误处理器

在生产环境的应用中，发生错误除了要显示错误信息给客户端之外，还需要上报错误、记录日志等操作，因此为了项目的可维护性，一般需要将错误处理中间件进行拆分，拆分为错误响应中间件、日志记录中间件等，每个中间件只负责一项工作。

```javascript
// 导入模块
const Koa = require('koa');

// 实例化应用
const app = new Koa();
```

```javascript
// 错误处理中间件
async function errorHandler(ctx, next) {
    try {
        await next();
    } catch (e) {
        ctx.status = e.status || 500;
        ctx.body = `System Error: ${e.message}`;
    }
}

// 错误日志记录中间件
async function errorLogger(ctx, next) {
    try {
        await next();
    } catch (e) {
        console.log(`${ctx.method} ${ctx.path} Error: ${e.message}`);
        throw e;          // 继续抛出错误,这样外层的 errorHandler 才能捕获到该错误
    }
}

app.use(errorHandler);    // 先挂载错误响应中间件
app.use(errorLogger);     // 后挂载错误日志中间件

// 抛出错误
app.use(ctx => {
    ctx.throw(403, 'Forbidden');
});

// 监听
app.listen(10000, () => {
    console.log('listen on 10000');
});
```

访问 http://localhost:10000,终端将显示"GET / Error: Forbidden",之后客户端将收到"System Error: Forbidden"。

5.6 路由系统

路由是一个 Web 应用的核心功能,当然,Koa 为了精简核心未包含路由功能,因此我们需要使用 koa-router 模块来实现路由功能。

5.6.1 快速开始

安装 koa-router 模块：

```
npm install koa-router -save
```

开始编码：

```javascript
// 导入模块
const Koa = require('koa');
const Router = require('koa-router');
// 实例化应用
const app = new Koa();
// 实例化路由
const router = new Router();

// 路由定义
router.get('/', async (ctx) => {
    ctx.body = 'Hello World';
});
router.get('/user', async (ctx) => {
    ctx.body = 'User';
});

// 挂载路由中间件
app.use(router.routes());
app.use(router.allowedMethods());

// 监听
app.listen(10000, () => {
    console.log('listen on 10000');
});
```

访问 http://localhost:10000，将输出"Hello World"；访问 http://localhost:10000/user，将输出"User"。

一定不要忘记通过 app.use()挂载路由中间件，否则路由将不会生效。

5.6.2 路由对象

路由需要实例化之后才能进行配置和挂载，路由的构造器函数签名如下：

```
function Router([options])
```

- options 选项，一般使用的选项就是 prefix。
  - prefix：路由前缀。

实例化后的路由示例和 Express 类似，路由定义的方法如下：

```
router.method(path, handler);
```

- method：HTTP 请求方法，支持 get/post/put/delete/head/options。
- path：路由路径。
- handler：路由处理函数，支持多个。

5.6.3 路由路径

koa-router 的路由路径支持字符串和字符串模式。

（1）以下是一些基于字符串的路由路径示例。

以下路由路径会将请求匹配到根路由/：

```
router.get('/', (ctx) => {
  ctx.body = 'Home';
});
```

以下路由路径会将请求匹配到/about：

```
router.get('/about', (ctx) => {
  ctx.body = '关于';
});
```

以下路由路径可以将请求匹配到/random.txt 文件：

```
router.get('/random.txt', (ctx) => {
  ctx.body = 'random.txt';
});
```

（2）以下是一些基于字符串模式的示例。

以下路由路径可以匹配/users/xxx，匹配成功后 xxx 将挂载到 ctx.params 变量下：

```
router.get('/users/:userId', (ctx) => {
  ctx.body = {
    userId: ctx.params.userId
  };
});
```

5.6.4 路由函数

koa-router 的路由函数和 Koa 默认的路由函数是相似的，也支持多个路由函数处理同一个请求。但是 ctx.params 只有 koa-router 的路由函数才可以访问。

```
router.get('/', async (ctx, next) => {
   ctx.state.data = { logged: true };
   await next();
}, (ctx) => {
  ctx.body = ctx.state.data;
});
```

5.6.5 路由级别中间件

Koa 默认的中间件是应用级别的，所有的请求都会被中间件处理。由于 koa-router 也支持多个路由函数，因此可以在指定路由或者整个路由对象上使用中间件。

以下是整个路由对象启用中间件的示例。

```
// 日志中间件
async function logger(ctx, next) {
   console.log(`${ctx.method} ${ctx.path} ${ctx.headers['user-agent']}`);
   await next();
}
router.use(logger);
router.get('/', (ctx) => {
   ctx.body = 'home';
});
```

```
// 日志中间件
async function logger(ctx, next) {
   console.log(`${ctx.method} ${ctx.path} ${ctx.headers['user-agent']}`);
   await next();
}

router.get('/', logger, (ctx) => {
   ctx.body = 'home';
});

router.get('/user', (ctx) => {
   ctx.body = 'user';
});
```

5.6.6 路由前缀

路由前缀可以将同一个模块的路由聚合在一起,提供一个统一的 URL 前缀供客户端访问。

```
const Koa = require('koa');
const Router = require('koa-router');

const app = new Koa();
const router = new Router({ prefix: '/user' });

router.get('/', (ctx) => {
   ctx.body = '/user';
});

router.get('/list', (ctx) => {
   ctx.body = 'user/list';
});
```

```
app.use(router.routes()).use(router.allowedMethods());
// 监听
app.listen(10000, () => {
    console.log('listen on 10000');
});
```

上面的代码将提供以下路由以供访问：

```
/user 输出 "/user"
/user/list 输出 "/user/list"
```

5.6.7 模块化路由

为了保持项目的可维护性，建议将路由逻辑拆分到其他模块中，否则入口 JS 将越来越大，不利于代码维护。

Koa 使用模块化路由的步骤非常简单：

- 独立文件中实现路由逻辑。
- 入口文件中挂载路由。

项目的目录结构如下：

```
|---koa.js
    |---routes
        |---site.js
        |---user.js
```

```
// routes/site.js
const Router = require('koa-router');
const router = new Router();

router.get('/', (ctx) => {
    ctx.body = '首页';
});

router.get('/about', (ctx) => {
    ctx.body = '关于页';
});

module.exports = router;
```

```
// routes/user.js
const Router = require('koa-router');
const router = new Router({ prefix: '/user' });

router.get('/', (ctx) => {
```

```js
    ctx.body = '用户首页';
});

router.get('/login', (ctx) => {
    ctx.body = '用户登录';
});

module.exports = router;
```

```js
// koa.js
// 导入模块
const Koa = require('koa');
const app = new Koa();

// 导入路由模块
const siteRoute = require('./routes/site');
const userRouter = require('./routes/user');

// 挂载路由
app.use(siteRoute.routes()).use(siteRoute.allowedMethods());
app.use(userRouter.routes()).use(siteRoute.allowedMethods());
// 监听
app.listen(10000, () => {
    console.log('listen on 10000');
});
```

以上例子将生成以下路由：

```
/user
/user/login
/
/about
```

5.7 模板渲染

Koa 也可以通过使用模板引擎来渲染 HTML 模板，为了减少学习门槛，本节还是使用 ejs 作为示例。

5.7.1 快速开始

使用 ejs 模板需要安装 koa-ejs 模块：

```
npm install koa-ejs --save
```

开始编码：

```
// koa.js
```

```
const Koa = require('koa');
const render = require('koa-ejs');
const app = new Koa();
render(app, {                        // 使用ejs中间件
    root: './templates',             // 模板目录
    layout: false,                   // 关闭模板布局
    viewExt: 'ejs'
});
app.use(async (ctx) => {
    ctx.state.name = 'xialei';       // ctx.state挂载数据
    await ctx.render('home', {
        now: (new Date()).toLocaleString(),
        title: '首页'
    });
});
// 监听
app.listen(10000, () => {
    console.log('listen on 10000');
});
```

```
<!-- templates/home.ejs -->
<p>欢迎你！<%= name %>! 现在时间：<%= now %></p>
```

访问 http://localhost:10000，将输出：

```
欢迎你！xialei! 现在时间：2019/11/10 下午19:25:55
```

Koa 渲染模板调用 ctx.render()方法，该方法签名如下：

```
ctx.render(view[, objects][, callback])
```

- view: 视图文件。
- objects: 视图变量。
- Callback: 回调函数。

koa-ejs 还支持 ctx.state，也就是说挂载到 ctx.state 中的变量可以直接在 ejs 模板中使用，不需要作为第二个参数传递给 ctx.render()。

5.7.2 模板布局

模板布局是一项很好用的功能，Web 页面一般是由以下结构组成的：

- 头部区域（包括导航栏）。
- 内容主体区域。
- 底部区域（包括版权声明、友情链接等）。

实际上头部区域和底部区域是一样的，真正有变化的内容是内容主体区域，因此可以将头部区域和底部区域作为不变的内容，内容主体作为变化内容来构建布局。

koa-ejs 启用模板布局需要配置 layout 选项，上节中配置为 false 关闭了布局，本节传递一个 main 作为参数，最终的布局模板文件名为 main.ejs。

```html
<!-- templates/main.ejs -->
<!DOCTYPE html>
<html lang="en">
<head>
    <meta charset="UTF-8">
    <meta name="viewport" content="width=device-width, initial-scale=1.0">
    <meta http-equiv="X-UA-Compatible" content="ie=edge">
    <title><%= title %></title>
</head>
<body>
    <header>我是头部</header>
    <%- body %>
    <footer>我是底部</footer>
</body>
</html>
```

可以看到头部和底部都是固定的，页面标题和 body 是通过变量控制的。

```html
<!-- templates/home.ejs -->
<p>欢迎你！<%= name %>！现在时间：<%= now %></p>
```

```js
// koa.js
const Koa = require('koa');
const render = require('koa-ejs');
const app = new Koa();
render(app, {
    root: './templates',      // 模板目录
    layout: 'main',           // 启用模板布局
    viewExt: 'ejs'
});
app.use(async (ctx) => {
    ctx.state.name = 'xialei';  // ctx.state 挂载数据
    await ctx.render('home', {
        now: (new Date()).toLocaleString(),
        title: '首页'
    });
});
// 监听
app.listen(10000, () => {
    console.log('listen on 10000');
});
```

启用模板布局后访问 http://localhost:10000/，输出如下：

```
我是头部
欢迎你！xialei！现在时间：2019/11/10 下午 19:28:55

我是底部
```

5.8 博客项目实战

拥有一个个人博客应该可以说是开发者刚入行时最急切的需求了，本节将和大家一起学习如何搭建个人博客。

在本项目中你将学到：

- 用户登录流程以及登录状态维护
- 使用中间件来保护受限资源访问

5.8.1 功能梳理

各位读者或多或少使用过别人提供的博客系统，总体来说，一个简单的博客一般有以下功能：

- 登录/注册（本项目不开放注册，用固定的账号密码）。
- 查看文章列表。
- 查看文章详情。
- 发表文章。
- 编辑文章。
- 删除文章。
- 项目结构。

本项目采用分层的思想构建，项目结构如下：

```
├── index.js
├── middlewares                中间件目录
│   └── authenticate.js        认证中间件
├── package.json
├── routes                     路由目录
│   ├── post.js                文章路由
│   ├── site.js                站点路由
│   └── user.js                用户路由
├── services                   业务目录
│   ├── post.js                文章业务
│   └── user.js                用户业务
└── templates                  模板目录
    ├── index.ejs              网站首页
    ├── login.ejs              登录表单
    ├── main.ejs               布局文件
    ├── post.ejs               文章详情
    ├── publish.ejs            发布表单
    └── update.ejs             编辑表单
```

该目录是笔者使用多年 Koa 经验总结的一套目录，根据职责划分，比较清晰，也比较容易维护。

5.8.2 项目代码

1. middleware/authenticate.js

认证中间件，负责解析 cookie，将登录状态挂载到 ctx.state 上，供后续使用。

```js
// 认证中间件
module.exports = async function (ctx, next) {
    const logged = ctx.cookies.get('logged', {signed: true});
    ctx.state.logged = !!logged;
    await next();
};
```

2. routes/post.js

文章相关路由，提供文章发布/编辑/详情/删除功能。

```js
const Router = require('koa-router');
const postService = require('../services/post');
const router = new Router();

// 发布表单页
router.get('/publish', async (ctx) => {
    await ctx.render('publish');
});

// 发布处理
router.post('/publish', async (ctx) => {
    const data = ctx.request.body;
    if (!data.title || !data.content) {
        ctx.throw(400, '您的请求有误');
    }
    const item = postService.publish(data.title, data.content);
    ctx.redirect(`/post/${item.id}`);
});

// 详情页
router.get('/post/:postId', async (ctx) => {
    const post = postService.show(ctx.params.postId);
    if (!post) {
        ctx.throw(404, '文章不存在');
    }
    await ctx.render('post', {
        post: post
    });
});

// 编辑表单页
```

```js
router.get('/update/:postId', async (ctx) => {
    const post = postService.show(ctx.params.postId);
    if (!post) {
        ctx.throw(404, '文章不存在');
    }
    await ctx.render('update', {
        post
    });
});

// 编辑处理
router.post('/update/:postId', async (ctx) => {
    const data = ctx.request.body;
    if (!data.title || !data.content) {
        ctx.throw(400, '您的请求有误');
    }
    const postId = ctx.params.postId;
    postService.update(postId, data.title, data.content);
    ctx.redirect(`/post/${postId}`);
});

// 删除
router.get('/delete/:postId', async (ctx) => {
    postService.delete(ctx.params.postId);
    ctx.redirect('/');
});

module.exports = router;
```

3. routes/site.js

网站首页，负责读取文章列表并渲染到 HTML 上。

```js
const Router = require('koa-router');
const postService = require('../services/post');

const router = new Router();

// 网站首页
router.get('/', async (ctx) => {
    const list = postService.list();
    await ctx.render('index', {
        list: list
    });
});

module.exports = router;
```

4. routes/user.js

用户相关路由，负责登录和退出登录。

```js
const Router = require('koa-router');
```

```js
const userService = require('../services/user');

const router = new Router();
// 登录表单页
router.get('/login', async (ctx) => {
    await ctx.render('login');
});
// 登录处理
router.post('/login', async (ctx) => {
    const data = ctx.request.body;
    if (!data.username || !data.password) {
        ctx.throw(400, '您的请求有误');
    }

    const logged = userService.login(data.username, data.password);
    if (!logged) {
        ctx.throw(400, '账号或密码错误');
    }
    ctx.cookies.set('logged', 1, {
        signed: true,
        httpOnly: true
    });
    ctx.redirect('/', '登录成功');
});
// 退出登录
router.get('/logout', (ctx) => {
    ctx.cookies.set('logged', 0, {
        maxAge: -1,
        signed: true
    });
    ctx.redirect('/', '退出登录成功');
});

module.exports = router;
```

5. services/post.js

文章相关业务，提供文章发表/删除/展示/编辑功能，由于还没有介绍数据库相关知识，本项目也是将数据保存到内存中，项目重启时数据会丢失。

采用独立的文件处理具体业务是很好的做法，比如后期我们学习完数据库来使用数据库改造本项目时，主要改动在 services/post.js，不需要更改其他地方。

```js
const fs = require('fs');
const bluebird = require('bluebird');
// Promise 化 fs 的 API
bluebird.promisifyAll(fs);

// 文章数据
const posts = [];
// 文章 ID
let postId = 1;
```

```js
// 发表文章
exports.publish = function (title, content) {
    const item = {
        id: postId++,
        title: title,
        content: content,
        time: (new Date()).toLocaleString()
    };
    posts.push(item);
    return item;
};

// 查看文章
exports.show = function (id) {
    id = Number(id);
    for (const post of posts) {
        if (post.id === id) {
            return post;
        }
    }
    return null;
};

// 编辑文章
exports.update = function (id, title, content) {
    id = Number(id);
    posts.forEach((post) => {
        if (post.id === id) {
            post.title = title;
            post.content = content;
        }
    });
};

// 删除文章
exports.delete = function (id) {
    id = Number(id);

    let index = -1;
    posts.forEach((post, i) => {
        if (post.id === id) {
            index = i;
        }
    });
    if (index > -1) {
        posts.splice(index, 1);
    }
};
```

```
// 文章列表
exports.list = function () {
    return posts.map(item => item);
};
```

koa-router 中 ctx.params.xxx 获取到的参数类型为字符串，post.js 文件中 id 是数字，需要转换一下。

6. services/user.js

用户业务，负责用户登录检测。账号密码是通过对象保存的，如果需要支持多个账号，直接给 user 对象添加属性即可。

```
const user = {
    xialei: 'password'
};
// 登录
exports.login = function (username, password) {
    if (user[username] === undefined) {
        return false;
    }
    return user[username] === password;
};
```

7. templates/index.ejs

网站首页模板，负责渲染文章列表。

```
<h1>文章列表</h1>
<% if(list.length === 0) { %>
    <p>暂时没有文章。</p>
<% } else { %>
    <table>
        <tr>
            <th>ID</th>
            <th>标题</th>
            <th>发表时间</th>
            <% if(logged) { %>
                <th>操作</th>
            <% } %>
        </tr>
        <% list.forEach((post) => { %>
            <td><%= post.id %></td>
            <td><a href="/post/<%= post.id %>"><%= post.title %></a></td>
            <td><%= post.time %></td>
            <% if(logged) { %>
                <th>
                    <a href="/update/<%= post.id %>">编辑</a>
                    <a href="/delete/<%= post.id %>" onclick="return confirm('确认删除吗?')">删除</a>
                </th>
            <% } %>
```

```
        <% }) %>
    </table>
<% } %>
```

8. templates/login.ejs

登录表单页,负责收集用户输入的数据并发送给服务器。

```html
<form action="/login" method="POST" enctype="application/x-www-form-urlencoded">
    <fieldset>
        <legend>登录</legend>
        <div>
            <label for="username">账号</label>
            <input type="text" name="username" id="username" required>
        </div>
        <div>
            <label for="password">密码</label>
            <input type="password" name="password" id="password" required>
        </div>
        <div>
            <button type="submit">登录</button>
        </div>
    </fieldset>
</form>
```

9. templates/main.ejs

主布局文件,根据登录状态来显示不同的顶部导航。

```html
<!DOCTYPE html>
<html lang="en">

<head>
    <meta charset="UTF-8">
    <meta name="viewport" content="width=device-width, initial-scale=1.0">
    <meta http-equiv="X-UA-Compatible" content="ie=edge">
    <title>博客</title>
</head>

<body>
<% if(logged) { %>
    <a href="/">首页</a>
    <a href="/publish">发表文章</a>
    <a href="/logout">退出登录</a>
<% } else { %>
    <a href="/login">登录</a>
<% } %>

<%- body %>
</body>

</html>
```

10. templates/post.ejs

文章详情页。

```
<div>
    <h1><%= post.title %></h1>
    <time>发表时间：<%= post.time %></time>
    <hr>
    <div><%= post.content %></div>
</div>
```

11. templates/publish.ejs

文章发布页，负责收集用户输入的数据并发送给服务器。

```
<form action="/publish" method="POST" enctype=
            "application/x-www-form-urlencoded">
    <fieldset>
        <legend>发表文章</legend>
        <div>
            <label for="title">标题</label>
            <input type="text" name="title" id="title" required>
        </div>
        <div>
            <label for="content">内容</label>
            <textarea name="content" id="content" required></textarea>
        </div>
        <div>
            <button type="submit">发布</button>
            <button type="reset">重置</button>
        </div>
    </fieldset>
</form>
```

12. templates/update.ejs

文章编辑页，负责填充当前文章的数据到输入框中，并收集用户提交的数据发送给服务器。

```
<form action="/update/<%= post.id %>" method="POST"
            enctype="application/x-www-form-urlencoded">
    <fieldset>
        <legend>编辑文章</legend>
        <div>
            <label for="title">标题</label>
            <input type="text" value="<%= post.title %>" name="title" id="title" required>
        </div>
        <div>
            <label for="content">内容</label>
            <textarea name="content" id="content" required><%= post.content %></textarea>
        </div>
        <div>
            <button type="submit">发布</button>
```

```
        <button type="reset">重置</button>
    </div>
  </fieldset>
</form>
```

13. index.js

入口 JS 文件，负责挂载中间件、路由、应用配置和启动服务器。

```
const Koa = require('koa');
const render = require('koa-ejs');
const bodyParser = require('koa-bodyparser');
const authenticate = require('./middlwares/authticate');
// 路由
const siteRoute = require('./routes/site');
const userRoute = require('./routes/user');
const postRoute = require('./routes/post');

const app = new Koa();
app.keys = ['hOOTTQctIjSjNykY'];
// 使用中间件
app.use(bodyParser());
app.use(authenticate);
render(app, {
    root: './templates',
    layout: 'main',
    viewExt: 'ejs'
});
// 挂载路由
app.use(siteRoute.routes()).use(siteRoute.allowedMethods());
app.use(userRoute.routes()).use(userRoute.allowedMethods());
app.use(postRoute.routes()).use(postRoute.allowedMethods());

app.listen(10000, () => {
    console.log('listen on 10000');
});
```

5.8.3 效果展示

运行项目之后可以访问 http://localhost:10000，页面效果说明如下。

首页（见图 5-3）

图 5-3

登录（见图5-4）

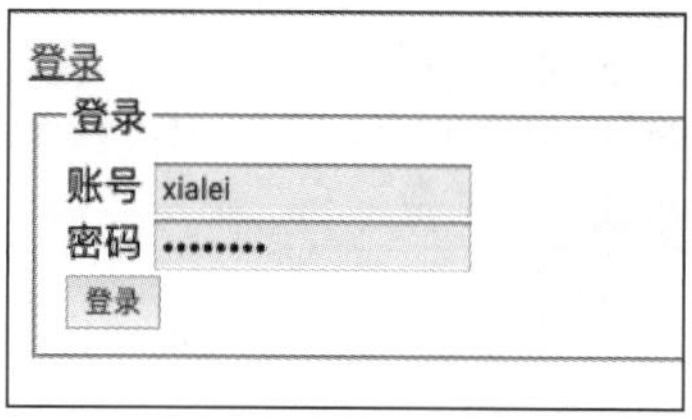

图 5-4

发表文章（见图5-5）

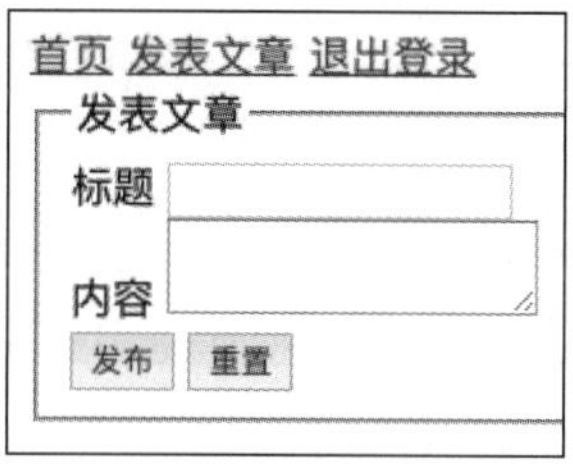

图 5-5

文章列表（首页有文章的情况，见图5-6）

图 5-6

文章详情（见图5-7）

图 5-7

文章编辑(见图 5-8)

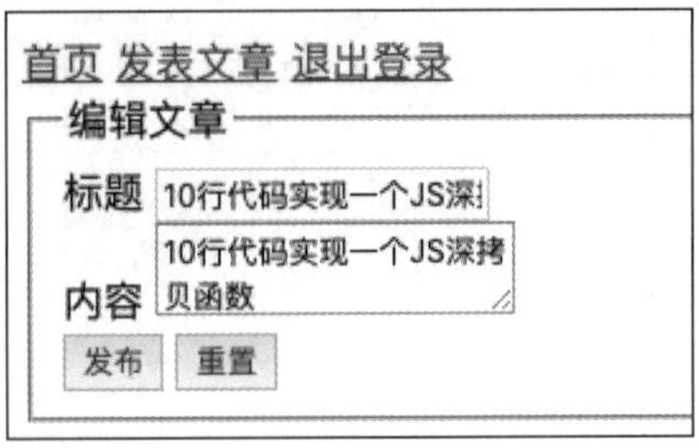

图 5-8

5.8.4 项目小结

本节和大家一起从零开发了一个简单的博客项目,项目采用了规范的文件结构,希望大家可以合理运用该结构进行项目开发。本项目主要使用到的技术有:

- 自定义 Koa 中间件。
- Koa 路由系统。
- Koa 模板渲染。
- Koa 表单处理。

5.9 本章小结

Koa 被称为"下一代的 Web 开发框架"确实有一定的道理,基于 async/await 设计的异步回调模型以及"洋葱圈"中间件模型是一个划时代的理念,让 Node.js 从回调地狱(Callback Hell)中彻底解放出来。

此外,由于极简的设计保障了 Koa 的运行性能和扩展性,因此一切都是可以按需扩展的。

回顾一下本章所学:

- Koa 的上下文对象、请求处理。
- Koa 的中间件的使用和编写。
- Koa 的路由系统。
- Koa 的模板渲染。
- 博客系统开发。

到目前为止,我们学习了主流的 Express 和 Koa 框架,在实际开发中,一般会结合其他软件(比如数据库、缓存等)来开发一个完整的应用。接下来我们将学习文档型 NoSQL 数据库 MongoDB。

第 6 章

文档型 NoSQL 数据库——MongoDB

本章内容

- MongoDB 的简介和安装
- MongoDB 的基本使用
- Node.js 操作 MongoDB

6.1 简　介

MongoDB 是 C++编写的、基于分布式文件存储的开源 NoSQL（Not Only SQL 的缩写，意思是"不仅仅是 SQL"）数据库系统。它是一个介于关系数据库和非关系数据库之间的产品，旨在为 Web 应用提供可扩展的高性能数据存储解决方案。

在高负载的情况下，通过添加更多的节点，可以保证服务器性能。

MongoDB 将数据存储为一个文档，数据结构由"键-值对"（Key-Value Pair）组成，是一种被称为"BeJSON"的类 JSON 数据。

```
{
  name: "test",
  status: "ok",
  "favorites": ["writing"]
}
```

6.1.1 主要特点

- MongoDB 是一个面向文档存储的数据库，操作起来比较简单。
- 通过增加节点可以提高服务器的性能（分片、分布式）。

- MongoDB 支持丰富的查询表达式，查询指令使用类似 JSON 的语法，可以查询内嵌文档中的对象和数组。
- MongoDB 是一个"最像"关系数据库的 NoSQL 数据库。

6.1.2 概念

不管学习何种数据库都需要明白其中的基础概念，MongoDB 的基础概念是文档、集合、数据库。表 6-1 所示对比传统关系型数据库进行介绍。

表 6-1　传统关系型数据库与 MongoDB 术语比较

SQL 术语	MongoDB 术语
database（数据库）	database（数据库）
table（数据表）	collection（集合）
row（数据行）	document（文档）
column（字段）	field（字段）
index（索引）	index（索引）

为了便于理解，我们对比一下同样的数据在 SQL 数据库和 MongoDB 数据库中存储的区别。

1. SQL 中的存储

id	username	email	created_at
1	demo	demo@demo.com	2000-01-01

2. MongoDB 中的存储

```
{
 "_id":ObjectId("5dcb9dab8efcaa346e94537a"),
 "username":"demo",
 "email":"demo@demo.com",
 "created_at":"2000-01-01"
}
```

6.1.3 数据库

MongoDB 的数据库可以看作是关系数据库中的数据库，一个服务器实例可以建立多个独立的数据库，每一个数据库都有自己的权限和数据，数据内容也存放在不同的文件中。

下面是 MongoDB 数据库的常用命令。

```
show dbs        # 查看数据库列表
db              # 查看当前数据库
use dbname      # 切换到指定的数据库
```

6.1.4 集合

MongoDB 的集合可以看作关系数据库中的表,用于存放一组数据,但是 MongoDB 和 SQL 数据库有一个最大的区别：

- 关系数据库的数据是结构化的，需要先定义表结构才能插入数据，要求每个数据行的结构一致。
- MongoDB 的数据是非结构化的，不需要定义表，每个数据（文档）之间的结构也不要求一致。

集合的命名规则：

- 集合名不能为空。
- 集合名不能含有 '\0' 字符，该字符表示字符串的结尾。
- 集合名不能以 "system." 开头，这是 MongoDB 保留使用的。

6.1.5 文档

MongoDB 的文档可以看作关系数据库中的数据行，但是 MongoDB 中同一个集合的文档结构不要求一致，也就是说下面两个文档是合法的。

```
> db.blog.find()
{ "_id" : ObjectId("5dcb9dab8efcaa346e94537a"), "title" : "test", "content" : "
    你好" }
{ "_id" : ObjectId("5dcba1a18efcaa346e94537b"), "name" : "xialei" }
```

需要注意的是：

- 文档中的 "键-值对" 是有序的。
- 文档支持丰富的数据类型，包含简单类型甚至嵌套文档。
- 文档区分数据类型和字母大小写。
- 文档不能有重复的键。

文档键名的规范：

- 键不能含有 '\0'。这个字符用来表示字符串的结尾。
- 不建议键名包含 '.' 和 '$'，这是 MongoDB 命令用到的字符。
- 不建议键名以下划线开头，这个是 MongoDB 保留的。

MongoDB 支持的数据类型：

- String：MongoDB 只支持 UTF-8 编码的字符串。
- Integer：根据服务器位数，可分为 32 位和 64 位整型。
- Boolean：存储布尔值 true/false。

- Double：双精度浮点值。
- Array：数组。
- Timestamp：时间戳，记录文档添加或修改的时间。
- Object：对象或嵌套文档。
- Null：空值。
- Date：日期时间。用 UNIX 时间格式来存储日期或时间，一般用 Timestamp 多一点。
- Object ID：文档的 ID。
- Binary Data：二进制数据。
- Code：JavaScript 代码。
- Regular Express：正则表达式。

比较常用的数据类型有 Boolean、String、Timestamp、Integer、Object、Array。

6.2 安装

MongoDB 为 Windows、Linux、macOS 操作系统分别提供了对应的二进制程序，我们直接使用即可。下面简单说明一下在这三个平台上安装和启动 MongoDB 数据库的过程。

6.2.1 Windows

新版的 MongoDB 只提供 x64 版本的二进制编译包，支持 Windows7+(含)以后的系统。
下载页面的网址为：https://www.mongodb.com/download-center/community。
最新版本的下载链接为：https://fastdl.mongodb.org/win32/mongodb-win32-x86_64-2012plus-4.2.1-signed.msi。下载界面如图 6-1 所示。

图 6-1

下载完成后以默认选项安装即可。

使用以下命令可以启动 MongoDB（如果数据目录不存在，则需要事先创建）。

```
mongod --dbpath c:\data\mongodb
```

使用以下命令可以连接运行中的 MongoDB：

```
mongo
```

连接 MongoDB 成功后会输出以下信息：

```
MongoDB shell version v4.2.1
>
```

输入 version()查看 MongoDB 的版本：

```
> version()
4.2.1
```

如果安装或者启动 MongoDB 服务遇到问题，可以在作者 GitHub 或者公众号进行反馈。

6.2.2 Linux

Linux 有众多发行版，如 Ubuntu、CentOS 等，可以通过源码编译安装 MongoDB，也可以使用 MongoDB 提供的二进制包，本节以 Ubuntu 为例。

下载页面的网址为：https://www.mongodb.com/download-center/community。

最新版本的下载链接为：https://repo.mongodb.org/apt/ubuntu/dists/bionic/mongodb-org/4.2/multiverse/binary-amd64/mongodb-org-server_4.2.1_amd64.deb。下载界面如图 6-2 所示。

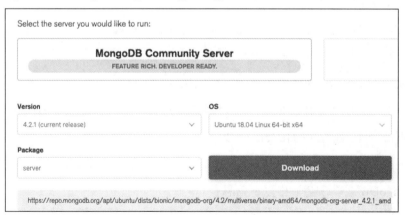

图 6-2

下载完二进制包之后，使用以下命令安装 MongoDB。

```
sudo dpkg -i mongodb-org-server_4.2.1_amd64.deb
```

在 Ubuntu 系统中安装完服务软件后一般会自动启动服务，如果需要手动启动 MongoDB 服务，可以执行以下命令：

```
service mongod start
```

使用以下命令可以连接运行中的 MongoDB：

```
mongo
```

连接 MongoDB 成功后会输出以下信息：

```
MongoDB shell version v4.2.1
>
```

输入 version() 查看 MongoDB 的版本：

```
> version()
4.2.1
```

Ubuntu 系统自带的软件仓库也有 MongoDB，只不过版本有点旧，笔者目前发现的 Ubuntu 仓库中的 MongoDB 版本为 3.6.3，不过作为学习使用的话，也可以直接安装，省去手动安装的问题。

使用以下命令直接安装系统软件仓库中的 MongoDB：

```
sudo apt-get install mongodb
```

6.2.3　macOS

MongoDB 提供给 macOS 的是编译好的二进制压缩包，包含二进制程序、配置等文件，需要进行一些配置才能使用，建议使用 Homebrew 安装 MongoDB。

```
brew tap mongodb/brew
brew install mongodb-community
```

安装完 MongoDB 之后可执行以下命令即可运行服务端程序：

```
mongod --config /usr/local/etc/mongod.conf
```

如果需要将 MongoDB 作为服务运行，可以执行以下命令：

```
brew services start mongodb/brew/mongodb-community
```

使用以下命令连接运行中的 MongoDB：

```
mongo
```

连接 MongoDB 成功后会输出以下信息：

```
MongoDB shell version v4.2.1
>
```

输入 version() 查看 MongoDB 的版本：

```
> version()
4.2.1
```

6.3 常用操作

使用 mongo 命令连接上 MongoDB 之后就可以进行数据操作了。下面我们学习一些 MongoDB 客户端常用的命令。

使用 MongoDB 客户端命令前需要先连接数据库：

```
mongo
```

6.3.1 创建数据库

MongoDB 创建数据库的语法如下：

```
use NAME
```

NAME 为数据库名称。如果数据库不存在，则自动创建；如果存在，则切换到该数据库。

下面是创建博客数据库的示例：

```
> use test
switched to db test
```

如果想查看所有数据库，可以使用以下命令：

```
> show dbs
admin   0.000GB
config  0.000GB
local   0.000GB
test    0.000GB
```

你可能会看不到刚才创建的数据库 blog，这是正常的。

在 MongoDB 中，如果数据库中没有内容是不会显示的，需要插入文档之后才会显示。

6.3.2 删除数据库

MongoDB 删除数据库的语法如下：

```
db.dropDatabase()
```

该语法只能删除当前使用的数据库，也就是 use 命令指定的数据库。

如果要删除指定数据库，则需要先使用 use 命令，然后再删除。

下面是删除博客数据库的示例：

```
> use blog
switched to db blog
> db.dropDatabase()
{ "ok" : 1 }
```

6.3.3 创建集合

MongoDB 创建集合的语法如下：

```
db.createCollection(name[, options])
```

- name：集合名称。
- options 选项：
  - capped：布尔型，如果为 true，则创建固定大小的集合（需要同时指定 size 选项），当达到最大值时，会自动覆盖最早的文档。
  - autoIndexId：布尔型，如果为 true，自动在 _id 字段建立索引。
  - size：整型，为固定大小的集合指定最大数据量，单位为 KB。
  - max：整型，为固定大小的集合指定最大文档数量。

固定集合一般用得比较少，本书不做研究。

下面是创建博客文章集合的示例。

```
> use blog
switched to db blog
> db.createCollection("post")
{ "ok" : 1 }
> db.post.insert({title:"你好", content:"Hello World"})
WriteResult({ "nInserted" : 1 })
> db.post.find()
{ "_id" : ObjectId("5dcbb37ca2fccdd5e21c6119"), "title" : "你好", "content" : "Hello
    World" }
```

ObjectId 是 MongoDB 自动生成的主键，每个文档都会有一个主键，不建议手动设置。

6.3.4 查看集合

MongoDB 查看集合的语法如下：

```
show collections
```

该命令只能查看当前数据库中的集合，如果需要查看其他数据库中的集合，需要使用 use 命令切换。

```
> show collections
post
```

6.3.5 删除集合

MongoDB 删除集合的语法如下：

```
db.COLLECTION.drop()
```

COLLECTION 为待删除的集合名称。该命令将删除当前数据库中的指定集合。

下面命令用于删除刚才创建的 post 集合示例：

```
> db.post.drop()
true
```

6.3.6 索引

索引能够极大地提高查询效率，如果不使用索引，MongoDB 必须遍历整个集合才能找出符合条件的文档。

索引类似于字典中的目录，当查询时若没有目录，我们就只能通过翻正文的方式去查找内容，而如果有了目录，则可以先翻目录得到页码，然后定位到对应的页面，由此可知查询效率的差别非常大。

MongoDB 创建索引的语法如下：

```
db.COLLECTION.createIndex(indexes[,options])
```

- indexes：字段设置，键为索引的文档键名，值为 1 时按升序创建索引，值为-1 时按降序创建索引。
- options 选项：
  - background：布尔型，设置 true 则在后台创建，服务依旧可用，传统关系数据库创建索引时会导致服务有一段时间不可用，默认为 false。
  - unique：布尔型，指定是否唯一索引。
  - name：字符串，索引名称。默认索引名称是字段名和排序顺序。

下面是给点赞字段创建索引的示例：

```
> db.post.createIndex({praise:-1})
{
  "createdCollectionAutomatically" : false,
  "numIndexesBefore" : 1,
  "numIndexesAfter" : 2,
  "ok" : 1
}
```

下面是其他索引命令：

- db.COLLECTION.getIndexes()：查看集合索引。
- db.COLLECTION.totalIndexSize()：查看索引占用的大小。
- db.COLLECTION.dropIndexes()：删除集合所有索引。
- db.COLLECTION.dropIndex(索引名称)：删除集合指定索引。

6.3.7 插入文档

MongoDB 插入文档的语法如下:

```
db.COLLECTION.insert(DOCUMENT)
```

- COLLECTION: 待插入文档的集合名称。
- DOCUMENT: 文档内容。

下面是向 post 集合中插入一篇文章的示例:

```
> use blog
switched to db blog
> db.createCollection("post")
> db.post.insert({title:"Node.js是什么", content:"Node.js是一个异步、非阻塞I/O的
    JavaScript运行环境",
... tags:["node.js"],
... meta:{
... praise:0,
... comment:0
... }
... }
... )
WriteResult({ "nInserted" : 1 })
```

MongoDB 命令在插入文档时可以一行一行输入,只要能正确关闭括号,插入就能成功。

```
> db.post.find()
{ "_id" : ObjectId("5dcbb7371250192a255f409f"), "title" : "Node.js是什么",
    "content" : "Node.js是一个异步、非阻塞I/O的JavaScript运行环境", "tags" :
    [ "node.js" ], "meta" : { "praise" : 0, "comment" : 0 } }
```

可以看到返回的数据和插入的数据一致,比如上面这些嵌套的数组也能正确展示,这就是 MongoDB 和传统关系数据库之间的差别。

6.3.8 更新文档

MongoDB 更新文档的语法如下:

```
db.COLLECTION.update(query, update[, options])
```

- COLLECTION: 待更新文档所在的集合。
- query: 查询对象。
- update: 更新数据。
- options: 选项。
  - ➢ upsert: 布尔型,如果未查询到需要更新的数据,是否将 update 参数作为新文档插入,默认值为 false。
  - ➢ multi: 布尔型,MongoDB 默认只更新第 1 条找到的文档,multi 为 true 时更新所有文档。

第 6 章 文档型 NoSQL 数据库——MongoDB

➢ writeConcert：抛出异常的级别。

以下是对刚才插入的"Node.js 是什么"的文档标题进行更新的示例：

```
> db.post.update({title:"Node.js是什么"},{title:"Node.js是什么【已更新】"})
WriteResult({ "nMatched" : 1, "nUpserted" : 0, "nModified" : 1 })
> db.post.find()
{ "_id" : ObjectId("5dcbb7371250192a255f409f"), "title" : "Node.js是什么【已更新】
" }
```

update 更新文档时，update 参数需要传入完整的文档属性，否则就会像上述例子一样发生文档数据丢失的问题！

6.3.9 删除文档

MongoDB 删除文档的语法如下：

```
db.COLLECTION.remove(query[, options])
```

- COLLECTION：待删除文档所在的集合。
- query：查询对象。
- options：选项。
  ➢ justOne：布尔型，设置为 true 时只删除符合条件的第 1 条文档，否则删除所有符合条件的文档，默认值为 false。
  ➢ writeConcert：抛出异常的级别。

以下是根据 ObjectId 来删除文章的示例：

```
> db.post.remove({_id:ObjectId("5dcbb7371250192a255f409f")})
WriteResult({ "nRemoved" : 1 })
```

6.3.10 查询文档

MongoDB 查询文档的语法如下：

```
db.COLLECTION.find(query[, projection])
```

- COLLECTION：待查询的集合。
- query：查询对象。
- projection：查询指定的键。默认返回所有键，设置该选项可以返回指定的键。

1．查询语法

下面我们先来学习 query 对象的使用方法。

MongoDB 支持丰富的查询语法，与 SQL 不同之处在于 MongoDB 需要使用规定的表达式查询，而 SQL 是偏向使用自然语言的查询语法。

表 6-2 列出了 MongoDB 的比较操作符。

表 6-2 MongoDB 的比较操作符

表达式	说明	示例
{key:value}	等于	db.post.find({title:"Node.js"})
{key:{$lt:value}}	小于	db.post.find({praise:{$lt:1}})
{key:{$lte:value}}	小于等于	db.post.find({praise:{$lte:1}})
{key:{$gt:value}}	大于	db.post.find({praise:{$gt:1}})
{key:{$gte:value}}	大于等于	db.post.find({praise:{$gte:1}})
{key:{$ne:value}}	不等于	db.post.find({praise:{$ne:1}})

多个表达式可以结合逻辑运算符完成更加复杂的查询，下面我们来看看 AND 和 OR 两个逻辑运算符。

（1）AND

多个查询条件是"与"的关系，也就是必须同时满足。

下面是查找标题为 Node.js 且点赞数大于 0 的示例：

```
db.post.find({
  title:"Node.js",
  praise:{$gt:0}
});
```

（2）OR

多个查询条件之间是"或"的关系，也就是最少满足一个条件即可。

下面是查找点赞数或评论数大于 0 的示例：

```
db.post.find({
  $or: [
    {praise:{$gt:0}},
    {comment:{$gt:0}}
  ]
})
```

2. projection

projection 在查询时可以指定返回哪些键或者屏蔽哪些键。

```
> db.post.find({})
{ "_id" : ObjectId("5dcbbbe81250192a255f40a0"), "title" : "HAHA" }
{ "_id" : ObjectId("5dcbbc971250192a255f40a1"), "title" : "文章1", "praise" : 0 }
{ "_id" : ObjectId("5dcbbc9a1250192a255f40a2"), "title" : "文章2", "praise" : 1 }
{ "_id" : ObjectId("5dcbbca01250192a255f40a3"), "title" : "文章3", "praise" : 2 }
```

我们选择只返回 title 键：

```
> db.post.find({},{title:1})
{ "_id" : ObjectId("5dcbbbe81250192a255f40a0"), "title" : "HAHA" }
{ "_id" : ObjectId("5dcbbc971250192a255f40a1"), "title" : "文章1" }
{ "_id" : ObjectId("5dcbbc9a1250192a255f40a2"), "title" : "文章2" }
```

```
{ "_id" : ObjectId("5dcbbca01250192a255f40a3"), "title" : "文章3" }
```

我们选择不返回 title，其他的键都返回：

```
> db.post.find({},{title:0})
{ "_id" : ObjectId("5dcbbbe81250192a255f40a0") }
{ "_id" : ObjectId("5dcbbc971250192a255f40a1"), "praise" : 0 }
{ "_id" : ObjectId("5dcbbc9a1250192a255f40a2"), "praise" : 1 }
{ "_id" : ObjectId("5dcbbca01250192a255f40a3"), "praise" : 2 }
```

可见 projection 对象的键就是需要操作的字段，值就是开关，设置为 1 就返回，设置为 0 就不返回。

6.3.11 其他查询语法

1. 分页查询

SQL 数据库通过 LIMIT 来返回指定区间内的数据行，MongoDB 通过 skip() 和 limit() 来实现该功能。

```
db.COLLECTION.find(query[,projection]).limit(LIMIT).skip(SKIP)
```

- LIMIT：限制返回的文档数量。
- SKIP：需要跳过的数量。

下面是一些示例：

```
db.post.find().limit(5)      # 返回前 5 条
db.post.find().skip(1)       # 跳过第 1 条
```

skip 和 limit 主要用来进行分页。

以下是一些分页的示例，分页大小为 10：

```
db.post.find().skip(0).limit(10)     # 返回第 1 页的 1 ~ 10 条数据
db.post.find().skip(10).limit(10)    # 返回第 2 页的 11 ~ 20 条数据
```

2. 统计文档数量

MongoDB 统计文档数量的语法如下：

```
db.COLLECITON.count(query)
```

- query：查询对象。

下面是统计点赞数大于 10 的示例：

```
db.post.count({
  praise:{$gt:10}
})
```

3. 排序

MongoDB 调用 sort() 方法来对数据进行排序，调用格式如下：

```
db.COLLECTION.find().sort(SORT)
```

- SORT：对象，键为字段，值为排序规则。1 为升序，-1 为降序。

下面是按照年龄从小到大进行排序的示例：

```
db.users.find().sort({age:1})
```

下面是按照年龄从大到小进行排序的示例：

```
db.users.find().sort({age:-1})
```

6.4 Node.js 集成

Node.js 一般使用 mongoose 模块来操作 MongoDB 数据库。下面针对常用的操作（Create、Read、Update、Delete，一般简称为 CRUD 操作）进行介绍。

6.4.1 初始化项目

使用 npm 初始化项目并安装 mongoose 模块：

```
mkdir mongoose-example
cd mongoose-example

npm init -y
npm install mongoose -save
```

6.4.2 连接数据库

mongoose 连接数据库的语法如下：

```
mongodb://[username:password@]host1[:port1][,host2[:port2],...[,hostN[:portN]]][/[database][?options]]' [, options]);
```

- username：连接账号。
- password：连接密码。
- host1: port1, host2:port2：数据库地址，Mongodb 支持分布式，因此可以传入多个服务器的地址。
- database：数据库名称。
- ?options：可选的 GET 形式的连接选项，例如?poolSize=4&replicaSet=test。
- options：可选的连接选项，常用选项如下：
  - useNewUrlParser：布尔型，使用新版本的 URL 解析器来解析连接字符串。
  - reconnectTries：数字，重连次数。

> reconnectInterval：数字，重连间隔，单位毫秒。
> poolSize：数字，连接池大小。

以下是连接本地 MongoDB 的 blog 数据库示例：

```
async function connect() {
   try {
      await mongoose.connect('mongodb://localhost:27017/blog',
   { useNewUrlParser: true });
      console.log('连接数据库成功');
   } catch (e) {
      console.log('连接数据库失败');
   }
}

connect();
```

6.4.3　mongoose 的关键概念

- Schema：模型类的骨架，通过 Schema 可以生成模型类，通过模型类可以生成文档。
- Model：由 Schema 产生的构造器，具有属性和行为。Model 的每一个实例就是一个 MongoDB 的 Document。
- Instance：Model 的实例，通过 new Model()得到。

6.4.4　Schema

每一个 Schema 对应 MongoDB 中的一个集合。Schema 定义了集合中文档的字段格式。

mongoose 出于可维护性和易用性的目的定义 Schema 来限定文档结构，但是 MongoDB 没有这个限制，也就是说 MongoDB 的文档无论什么结构都可以存储。

1. 定义 Schema

定义 Schema 的语法有以下两种（都是可用的）：

```
const schema = new Schema({
 字段名:字段类型
});

const schema = new Schema({
 字段名：选项
});
```

字段类型支持以下 JS 的标准类型：

- String
- Boolean
- Number

- Date
- Object
- Array

选项是高度自定义的方式，可以定义字段必填、默认值等属性，支持的属性如下：

- type：数据类型。
- default：默认值。
- index：索引选项。
  - background：是否后台创建。
  - unique：是否唯一索引。
- required：是否必填。
- unique：是否唯一索引。
- min：最小值，Number 类型字段可用。
- max：最大值，Number 类型字段可用。

以下是定义博客文章的 Schema 示例：

```
const Post = new Schema({
   title: { type: String, required: true, unique: true },
   content: { type: String, required: true },
   created_at: { type: Date, default: new Date(), index: { background: false, unique:
    false } },
   tags: [String],
   published: Boolean,
   meta: {
       praise: Number,
       comment: Number,
   }
});
```

我们知道 tags 是数组，因此需要定义为数组类型[String]，String 是限定数组内的数据类型为字符串。

meta 是普通的 Object，该 Object 有两个字段。

2. 定义实例方法

前面的内容中提到，Schema 构建出 Model，Model 实例化之后得到文档实例。如果需要给每个文档实例添加一些方法进行调用，可以将其定义在 Schema 上。

Schema 定义的实例方法最终会挂载到 Model.prototype 对象上，所有 Model 的实例都可以调用。

定义实例方法的语法如下：

```
schema.methods.方法名 = function(方法参数) {
  // 业务代码
}
```

下面是在刚才的 Schema 中添加获取点赞数量的实例方法示例。

```
const Post = new Schema({
    title: { type: String, required: true, unique: true },
    content: { type: String, required: true },
    created_at: { type: Date, default: new Date(), index: { background: false,
    unique: false } },
    tags: [String],
    published: Boolean,
    meta: {
        praise: Number,
        comment: Number,
    }
});

// 定义实例方法
Post.methods.getPraise = function () {
    return this.meta.praise;
}

const Model = mongoose.model('Post', Post);
const news = new Model();
news.meta = {praise: 0, comment: 0};
console.log(news.getPraise()); // 调用实例方法
```

3. 定义静态方法

Schema 还支持定义静态方法，构建为 Model 后直接通过 Model 调用即可。

定义静态方法的语法如下：

```
schema.statics.方法名 = function(方法参数) {
  // 业务代码
}
```

以下是刚才的 Schema 添加构造模型的静态方法示例。

```
const Post = new Schema({
   title: { type: String, required: true, unique: true },
   content: { type: String, required: true },
   created_at: { type: Date, default: new Date(), index: { background: false, unique:
    false } },
   tags: [String],
   published: Boolean,
   meta: {
      praise: Number,
      comment: Number,
   }
});
// 定义静态方法
Post.statics.new = function (data) {
   return new Model(data);
```

```
}
const Model = mongoose.model('Post', Post);
const data = Model.new();    // 调用静态方法
console.log(data);
```

6.4.5　Model

Model 通过 Schema 构建而来，Schema 定义了文档的骨架，Model 用来和数据库交互。

1．插入数据

插入单条数据有以下几种写法。

（1）实例化 Model 时传递文档，并调用 save()完成保存。

```
const Model = mongoose.model('Post', Post);
const news = new Model({
    title: 'Node.js',
    content: 'Node.js 内容'
});
await news.save();
```

（2）实例化 Model 时不传递文档，直接对实例赋值，并调用 save()完成保存。

```
const Model = mongoose.model('Post', Post);
const news = new Model();
news.title = 'Node.js 是什么？';
news.content = 'Node.js 是什么？';
await news.save();
直接通过 Model.create()完成保存。

await Model.create({
    title: 'Node.js 示例',
    content: 'Node.js 示例内容'
});
```

插入多条数据有以下几种写法。

（1）调用 Model.insertMany()并传递数组完成插入。

```
await Model.insertMany([{
    title: 'Node.js 示例 1',
    content: 'Node.js 示例内容 1'
}, {
    title: 'Node.js 示例 2',
    content: 'Node.js 示例内容 2'
}]);
```

（2）调用 Model.create()并传递数组完成插入（Model.create()同时支持单条和多条）。

```
await Model.create([{
```

```
    title: 'Node.js 示例1',
    content: 'Node.js 示例内容1'
}, {
    title: 'Node.js 示例2',
    content: 'Node.js 示例内容2'
}]);
```

2. 查询数据

调用 Model.findOne(query)查询满足条件的第 1 条数据：

```
const news = await Model.findOne({ title: 'Node.js 示例2' });
```

调用 Model.findById()查询指定 ID 的数据：

```
const news = await Model.findById('5dcbd9eadf602e9196d47fa7');
```

调用 Model.find(query)查询多条数据：

```
const Model = mongoose.model('Post', Post);
const news = await Model.find().skip(5).limit(5);
```

其他查询语法和 MongoDB 的查询语法类似，这里不再赘述。

3. 更新文档

调用 Model.updateOne(query, update)更新符合条件的一个文档：

```
await Model.updateOne({ _id: '5dcbd9eadf602e9196d47fa7' }, { title: 'Node.js 示
    例2【已修改】' });
```

该方法更新文档是增量更新，也就是说只修改指定的字段，剩余的字段数据依旧保存。而 MongoDB 命令编译文档时需要提供完整的更新文档。

调用 Model.updateMany(query, update)更新符合条件的所有文档：

```
await Model.updateMany({ published: true }, { title: 'Node.js 示例2【已修改】' });
```

4. 删除文档

调用 Model.deleteOne(query)删除一个符合条件的文档：

```
await Model.deleteOne({ _id: '5dcbd9eadf602e9196d47fa7' });
```

调用 Model.deleteMany(query)来删除所有符合条件的文档：

```
await Model.deleteMany({ praise: 0 });
```

6.5 本章小结

由于 MongoDB 的文档结构跟 JSON 很相似，天然拥有 JS 语言的亲和性，因此使用 Node.js 操作 MongoDB 变得简单。MongoDB 用来存放一些非结构化数据的优势很明显，在非结构化

数据高速发展的今天，起着非常大的作用。

回顾一下本章所学：

- MongoDB 简介。
- MongoDB 安装。
- MongoDB 基本概念和基本命令。
- Node.js 使用 mongoose 来操作 MongoDB。

下一章将学习目前最流行的关系数据库——MySQL，MySQL 有着完整的事务支持，可以用来存储一些重要数据。

第 7 章

最流行的关系型数据库——MySQL

本章内容

- MySQL 简介和安装
- MySQL 的基本语法
- MySQL 的关联关系
- MySQL 的事务操作

7.1 简　介

MySQL 是一个关系型数据库管理系统，目前属于 Oracle 旗下的产品。在 Web 应用方面 MySQL 是最好的 RDBMS（Relational Database Management System，关系型数据库管理系统）。

MySQL 主要特点：

- 开源、免费，因此不需要为此支付费用（社区版本）。
- MySQL 可以处理拥有上千万条记录的数据表。
- 使用标准的 SQL 语言。
- 支持多种操作系统和多种客户端语言。

7.2 安　装

MySQL 为支持的平台定期发布二进制软件包，下载页面的网址为 https://dev.mysql.com

/downloads/mysql/。

7.2.1 Windows

Windows 推荐使用 MySQL Installer 进行安装，MySQL Installer 可以安装服务器、客户端和管理工具等软件，截至发稿前最新版本的 MySQL Installer 为 8.0.18。

下载地址如下：

https://dev.mysql.com/get/Downloads/MySQLInstaller/mysql-installer-web-community-8.0.18.0.msi

下载后依照提示安装即可。

7.2.2 Linux

Linux 主流发行版为 Ubuntu 和 CentOS，这些发行版的软件仓库都包含了 MySQL 相关软件。

1. Ubuntu

如果你使用的是 Ubuntu 操作系统，可以使用下面的命令安装 MySQL：

```
sudo apt-get update
sudo apt-get install mysql-server mysql-client
```

安装完成后服务会自动启动，如果需要手动管理服务，可以使用下面的命令：

```
service mysql {start|stop|restart}
```

MySQL 服务初次启动时需要进行安全配置，服务运行后使用以下命令进行安全配置：

```
sudo mysql_secure_installation
```

执行成功后进入配置流程，MySQL 会循环一系列问题来完成初始化配置：

```
是否启用密码验证插件？
VALIDATE PASSWORD PLUGIN can be used to test passwords...
Press y|Y for Yes, any other key for No: N (不启用密码验证插件)
设置root账户密码(root账户是MySQL中权限最高的账户)

Please set the password for root here...
New password: (输入密码)
Re-enter new password: (重复输入)

是否移除匿名账户？
By default, a MySQL installation has an anonymous user,
allowing anyone to log into MySQL without having to have
a user account created for them...
Remove anonymous users? (Press y|Y for Yes, any other key for No) : Y (移除匿名账户)

是否禁用root账户远程连接？(禁用后root账户只能通过localhost连接数据库)
```

```
Normally, root should only be allowed to connect from
'localhost'. This ensures that someone cannot guess at
the root password from the network...
Disallow root login remotely? (Press y|Y for Yes, any other key for No) : Y (禁
    用 root 远程连接)
```

是否移除测试数据库?
```
By default, MySQL comes with a database named 'test' that
anyone can access...
Remove test database and access to it? (Press y|Y for Yes, any other key for No) :
    N (保留测试数据库)
```

是否重新加载权限表?
```
Reloading the privilege tables will ensure that all changes
made so far will take effect immediately.
Reload privilege tables now? (Press y|Y for Yes, any other key for No) : Y (重新
    加载权限表)
```

Ubuntu 操作系统下的 MySQL 安装到此结束。

2. CentOS

如果你使用的是 CentOS 操作系统，可以使用下面的步骤安装 MySQL。

下载并安装 MySQL 官方的 Yum Repository（Yum 软件仓库）：

```
wget http://dev.mysql.com/get/mysql57-community-release-el7-10.noarch.rpm
yum -y install mysql57-community-release-el7-10.noarch.rpm
```

使用 Yum 安装 MySQL 服务器：

```
yum -y install mysql-community-server
```

安装完成后服务会自动启动，如果需要手动管理服务，可以使用下面的命令：

```
service mysqld {start|stop|restart}
```

MySQL 服务正常运行，但是登录 MySQL 还需要 root 账号和密码，CentOS 安装 MySQL 后的处理和 Ubuntu 不一致，以下是处理过程：

使用以下命令查找临时密码：

```
grep 'temporary password' /var/log/mysqld.log
```

登录数据库。该命令会提示输入密码，输入刚才的临时密码之后回车键即可，终端不会显示密码，这是正常的：

```
mysql -uroot -p
```

修改默认密码：

```
mysql> ALTER USER 'root'@'localhost' IDENTIFIED BY '新密码';
```

重载权限表：

```
mysql> flush privileges;
```

退出 MySQL 客户端：

```
mysql>exit
```

CentOS 操作系统下的 MySQL 安装到此结束。

7.2.3 macOS

MySQL 官方提供了 macOS 的 DMG 安装包，macOS 下载之后即可完成安装。下载链接为：https://dev.mysql.com/get/Downloads/MySQL-8.0/mysql-8.0.18-macos10.14-x86_64.dmg。

另外一种方式是使用 Homebrew 来安装，这也是笔者建议的方式。下面是 Homebrew 安装 MySQL 的步骤。

安装 MySQL 服务器：

```
brew install mysql
```

安装 MySQL 客户端：

```
brew install mysql-client
```

启动 MySQL 服务：

```
brew services start mysql
```

macOS 下的 MySQL 安装后也需要进行初始化设置，执行以下命令进入初始化设置，设置步骤同 Ubuntu 系统：

```
mysql_secure_installation
```

macOS 操作系统下的 MySQL 安装到此结束。

安装或者管理 MySQL 服务的过程中遇到问题可以在作者 GitHub 或者公众号提问。

7.3 术　语

在学习 MySQL 数据库之前，需要先了解 RDBMS 的一些通用术语，如表 7-1 所示。

表 7-1　RDBMS 的一些通用术语

术语	说明
数据库	一些关联表的集合
数据表	一些相同结构的数据行的集合，如 10 个用户数据行组成的表
数据行	一条数据，如一个用户的数据
数据列	数据表中相同字段的数据，如用户表的账号列

（续表）

术语	说明
主键	用来标识记录的字段，同一个表中具有唯一性
外键	用于关联并约束两个表的关系
索引	对数据表一列或多列的值排好序的结构，类似于书籍的目录

7.4 索　引

索引的功能类似于字典的目录，能够快速地检索所需的数据。在不使用索引的情况下，我们查找一个字需要到正文中进行查找；而使用索引可以先查询目录再根据页码查询中文，这样可以减少查询时间。

索引的优点是可以加快数据的查询速度，当然缺点也有：在修改数据时，需要同时更新索引，所以索引会降低数据修改的速度。另外，索引也需要占用磁盘空间，所以索引不是越多越好，只有在必要的时候建立合适的索引才是推荐的做法。

7.4.1　普通索引

普通索引是最基本的索引，使用的时候没有任何限制。可以在建表时就指定好，也可以给已存在的表添加索引。

```
# 建表时指定索引
CREATE TABLE table_name (
  id INT NOT NULL,
  username VARCHAR(16) NOT NULL,
  PRIMARY KEY(id),
  KEY username(username)
);
# 已存在的表添加索引
ALTER TABLE `user` ADD INDEX `username`(`username`);
# 删除已存在的索引
ALERT TABLE `user` DROP INDEX `username`;
```

7.4.2　唯一索引

唯一索引和普通索引类似，唯一不同的是：唯一索引列的值必须唯一，但允许为 NULL。如果是组合索引，则列值的组合必须唯一。它有以下几种创建方式：

```
# 建表时指定
CREATE TABLE table_name (
  id INT NOT NULL,
```

```
  username VARCHAR(16) NOT NULL,
  PRIMARY KEY(id),
  UNIQUE KEY username(username)
);
# 已存在的表添加索引
ALTER TABLE `user` ADD UNIQUE INDEX `username`(`username`);
```

7.4.3 联合索引

之前介绍的索引都是一个字段的，如果需要使用多个字段进行筛选或者标识唯一性，需要使用联合索引，也就是多个字段构成的索引。联合索引也可以在建表时指定，或者给已存在的表添加索引。

```
# 建表时指定
CREATE TABLE `student_lesson` (
  `id` INT unsigned zerofill NOT NULL AUTO_INCREMENT COMMENT 'ID',
  `student_id` INT NOT NULL,
  `lesson_id` INT NOT NULL,
  PRIMARY KEY (`id`),
  UNIQUE KEY `uniq_student_lesson` (`student_id`,`lesson_id`)
) ENGINE=InnoDB AUTO_INCREMENT=2 DEFAULT CHARSET=utf8mb4;
# 已存在的表添加索引
ALTER TABLE `student_lesson` ADD UNIQUE INDEX `uniq_student_lesson`(`student_id`,
    `lesson_id`);
```

7.5 事　务

事务是由一组 SQL 语句组成的逻辑处理单元，要么全部执行成功，要么全部执行不成功。

例如银行汇款的例子，先扣除转出方金额，再增加转入方金额，这两个步骤要么全部执行成功，也就是汇款成功，钱从转出方到了转入方；要么全部执行不成功，也就是钱没有从转出方到转出方，这些是由事务来保证的。

如果没有事务或者不支持事务，那么假设在扣除转出方金额时，服务器出现故障，此时并没有把钱转给转入方，此时数据就出现不一致了，转出方的钱就消失了。

MySQL 中使用 InnoDB 存储引擎才能使用事务特性。MySQL 将服务器层与存储引擎进行了拆分，存储引擎负责数据的存储、查询等操作，服务器层进行安全验证、SQL 优化等通用操作。

7.5.1 ACID 原则

一般来说，事务必须满足 4 个原则（ACID）：原子性（Atomicity）、一致性（Consistency）、隔离性（Isolation）、持久性（Durability）。

- 原子性：一个事务中的操作，要么全部不完成，不会结束在中间某个环节。事务在执行过程中发生错误，会被回滚（Rollback）到事务开始前的状态，就像这个事务从来没有执行过一样。
- 一致性：在事务开始之前和事务结束以后，数据库的完整性没有被破坏。比如银行汇款的例子中，转出之前和转出之后的总金额不能有变化。
- 隔离性：数据库允许多个事务同时执行，隔离性可以防止多个事务并发执行时导致数据的不一致。事务隔离分为不同级别，包括读未提交（Read Uncommitted）、读已提交（Read Committed）、可重复读（Repeatable Read）和串行化（Serializable）。
- 持久性：事务提交后，对数据的修改就是永久的，即使服务器出现故障也不会丢失。

7.5.2 事务并发问题

事务并发执行并不是安全的，通常会发生以下问题：

- 脏读：事务 A 读取了事务 B 更新的数据，之后事务 B 回滚，此时 A 读取到的数据就是脏数据。
- 不可重复读：事务 A 多次读取同一数据，事务 B 在事务 A 多次读取的过程中，对数据做了更新并提交，导致事务 A 多次读取到的数据不一致。
- 幻读：事务 A 批量编辑数据，此时事务 B 插入了一条新的记录并提交，在事务 A 提交后发现还有一条数据并没有发生更新，就好像出现幻觉一样。

7.5.3 隔离级别

MySQL 支持四种隔离级别，这些隔离级别如下：

- 读未提交。当前事务都可以读取到其他事务未提交的数据。读未提交会引发脏读问题，该隔离级别很少使用。
- 读已提交。当前事务只能读取到已经结束事务后的数据。读已提交解决了脏读问题，但是会引发不可重复读问题。
- 可重复读。多个事务并发读取数据时，会看到同样的数据行。可重复读解决了不可重复读问题，但是会引发幻读问题，这是 MySQL 默认的事务隔离级别。
- 串行化。通过强制事务排序，使之不可能相互冲突，从而解决幻读问题。串行化的性能是最差的。

7.5.4 事务控制语句

默认情况下，MySQL 的 SQL 执行完毕后会自动提交，可以通过 SET AUTOCOMMIT=0 来关闭自动提交，不过一般情况下不建议关闭该选项。MySQL 提供了一系列手动控制事务的语句：

- BEGIN：开启一个事务。
- COMMIT：提交一个事务。
- ROLLBACK：回滚一个事务。
- SET TRANSACTION：设置事务隔离级别。MySQL 的 InnoDB 存储引擎支持 READ UNCOMMITTED（读未提交）、READ COMMITTED（读已提交）、REPEATABLE READ（可重复读）和 SERIALIZABLE（串行化）四种级别。

接下来我们用 MySQL 事务来完成银行汇款的操作：

```
BEGIN;    # 开启事务
UPDATE `account` SET balance=balance-100 WHERE name='转出方';
UPDATE `account` SET balance=balance+100 WHERE name='转入方';
COMMIT; # 提交事务
```

事务并不是没有开销的，事务的运行会涉及锁数据行甚至锁数据表，因此只有在必要的情况下才需要使用事务。根据经验，一般涉及对两个或两个以上的表进行数据修改（添加、修改、删除）时需要用到事务。

7.6 关联关系

在实际开发中，需要根据实体的内容设计数据表，实体间会有各种关联关系。MySQL 数据库中关联关系有 3 种：一对多关联、一对一关联与多对多关联。

7.6.1 一对多关联

一对多关联是最常见的一种关联关系。比如微博与用户的关系，一个用户会有很多微博，而一篇微博不能属于多个用户。也就是说用户表的一行在微博表会有多个匹配行，但在微博表的一行在用户表只会有一个匹配行。

一对多关联需要在数据多的表添加字段保存所属主体的 ID。

1. 用户表

Id	username	password	created_at
1	demo	-	2019-10-01 10:00:00

2. 微博表

id	content	created_at	user_id
1	微博 1	2019-10-01 11:00:00	1
2	微博 2	2019-10-01 12:00:00	2

本例中在微博表添加了 user_id 字段来存储所属用户的 ID。

7.6.2 一对一关联

一对一关联在实际生活中比较常见，比如一个人最多有一张身份证，也就是说用户表的一行在身份证表最多有一行记录，允许没有（比如没办身份证），身份证表的一行在用户表也只能有一行记录（一个身份证不能给两个人用）。

一对一关联需要分清主从，从表中的数据是不可以独立存在的，人可以没有身份证，但是身份证一定是某个人的。因此人是主，身份证是从。

一对一关联中，从表一般直接使用主表的主键作为标识键。

1. 人表

id	username	password	created_at
1	Zhangsan	-	2019-10-01 10:00:00

2. 身份证表

user_id	realname	sex	cardno	created_at
1	张三	1	xxx	2019-10-01 11:00:00

本例中身份证表直接使用了 user 表的主键作为外键存储所属人的信息。

7.6.3 多对多关联

多对多关联也是数据表的一种关联关系，比如学生和课程的关系，一个学生可以选择多门课程，一门课程也可以有多个学生选择，也就是说学生表的一行在课程表有多个匹配行，课程表的一行在学生表也有多个匹配行。

多对多关联一般需要额外添加一张关联表来记录关联关系，该表会存储两个外键，分别是学生 ID 和课程 ID。在多对多关联表中，学生 ID 可以重复，课程 ID 也可以重复，但是对于同一个学生 ID 和课程 ID 不可以重复，也就是同一个学生不能重复选择一门课程。

1. 学生表

id	name	created_at
1	张三	2019-10-01 10:00:00
2	李四	2019-10-01 10:00:00

2. 课程表

id	name	created_at
1	数学	2019-10-01 10:00:00
2	语文	2019-10-01 10:00:00

3. 选课表

id	学生 ID	课程 ID	created_at
1	1	1	2019-10-01 10:00:00
2	1	2	2019-10-01 10:00:00
3	2	1	2019-10-01 10:00:00

本例中张三选择了数学和语文课，李四选择了数学课。

7.7 数据库操作

使用 MySQL 命令之前需要先连接数据库。鉴于目前是学习阶段，直接使用 root 账号连接即可：

```
mysql -uroot -p
```

根据提示输入密码之后按回车键确认，如果遇到下面的问题请检查账号密码是否正确：

```
ERROR 1045 (28000): Access denied for user 'root'@'localhost' (using password: YES)
```

1. 查看数据库列表

在 MySQL 中命令不区分字母大小写，命令需要使用分号结尾才会执行：

```
mysql> show databases;
+--------------------+
| Database           |
+--------------------+
| information_schema |
| mysql              |
| performance_schema |
| sys                |
| test               |
+--------------------+
```

2. 选择数据库

在连接到 MySQL 服务器后，有多个数据库可供操作，所以需要选择要操作的数据库：

```
mysql> use test; # 使用 test 数据库
Database changed
```

SQL 命令不区分字母大小写，但是数据库名、表名、字段名区分字母大小写。

3. 创建数据库

```
mysql> create database demo;
Query OK, 1 row affected (0.01 sec)
```

MySQL 数据库命名规范和注意事项：

- 仅使用 26 个英文字母（区分大小写）和 0~9 这十个数字以及下画线 "_"，共 62 个字符。
- 数据库名称不得超过 64 个字符。
- 尽量不要使用 MySQL 关键字作为数据库名称。

4. 删除数据库

```
mysql> drop database demo;
Query OK, 0 rows affected (0.02 sec)
```

请谨慎执行数据库删除命令！MySQL 在删除数据库的时候没有额外的确认操作，因此容易造成误删，需要小心执行。

7.8 数据类型

在介绍数据表相关操作之前，需要介绍一下 MySQL 支持的数据类型。MySQL 中选择合适的数据类型对于数据库优化是十分重要的。

MySQL 支持丰富的数据类型，大致可以分为数值、字符串和日期/时间三类。

1. 数值类型

MySQL 支持 SQL 标准中定义的所有数值数据类型。SQL 标准中数值类型包括：

- 严格数值数据类型：INTEGER、SMALLINT、DECIMAL、NUMERIC。
- 近似数值数据类型：FLOAT、REAL、DOUBLE、PRECISION。

除了标准的数值类型之外，MySQL 还支持 TINYINT、MEDIUMINT、BIGINT。

表 7-2 是 MySQL 中整数类型的大小和数据范围。

表 7-2 MySQL 中整数类型的大小和数据范围

类型	大小	范围（有符号）	范围（无符号）
TINYINT	1 字节	$(-2^7, 2^7-1)$	$(0, 2^8-1)$
SMALLINT	2 字节	$(-2^{15}, 2^{15}-1)$	$(0, 2^{16}-1)$
MEDIUMINT	3 字节	$(-2^{23}, 2^{23}-1)$	$(0, 2^{24}-1)$
INT/INTEGER	4 字节	$(-2^{31}, -2^{31}-1)$	$(0，2^{32}-1)$
BIGINT	8 字节	$(-2^{63}, 2^{63}-1)$	$(0, 2^{64}-1)$

下面是 MySQL 中小数类型的大小和范围。

- FLOAT 4 字节（近似值）：
  - 有符号：-3.402823466E+38 ~ -1.175494351E-38, 0, 1.175494351E-38 ~ 3.402823466E+38
  - 无符号：0, 1.175494351E-38 ~ 3.402823466E+38
- DOUBLE 8 字节（近似值）：
  - 有符号：-1.7976931348623157E+308 ~ -2.2250738585072014E-308, 0, 2.2250738585072014E-308 ~ 1.7976931348623157E+308
  - 无符号：0, 2.2250738585072014E-308 ~ 1.7976931348623157E+308
- DECIMAL（精确值）：
  定义格式为 DECIMAL(P,D)，P 为有效数字的精度，范围 1~65，D 为小数位数，范围 0~30。MySQL 要求 D <= P。如果 P>D，则大小为 M+2 字节，否则为 D+2 字节。

2. 日期和时间类型

MySQL 中表示日期和时间的数据类型有 DATETIME、DATE、TIMESTAMP、TIME、YEAR。

每个日期和时间类型都有零值，如 DATE 类型的"0000-00-00"，当传入不合法的数据时 MySQL 将使用零值或报错（取决于 MySQL 设置）。

表 7-3 是 MySQL 中日期和时间类型的大小和数据范围。

表 7-3 MySQL 中日期和时间类型的大小和数据范围

类型	大小	范围	备注
DATE	3	1000-01-01 ~ 9999-12-31	
TIME	3	-838:59:59 ~ -838:59:59	时刻或持续时长
YEAR	1	901 ~ 2155	
DATETIME	8	1000-01-01 00:00:00 ~ 9999-12-31 23:59:59	
TIMESTAMP	4	1970-01-01 00:00:00 ~ 2038-01-19 03:14:07	

TIMESTAMP 类型支持自动更新，也就是数据插入或编辑时由 MySQL 来更新该类型字段的数据，不需要应用程序处理。

MySQL 中表示字符串的数据类型有 CHAR、VARCHAR、TINYBLOB、TINYTEXT、BLOB、TEXT、MEDIUMBLOB、MEDIUMTEXT、LONGBLOG、LONGTEXT。

表 7-4 是 MySQL 中字符串类型的大小和范围数据。

表 7-4　MySQL 中字符串类型的大小和范围数据

类型	大小	说明
CHAR	0～255 字符	定长字符串
VARCHAR	0～65535 字节	变长字符串
TINYBLOB	0～255 字节	短二进制数据
TINYTEXT	0～255 字节	短文本
BLOB	0～65535 字节	长二进制数据
TEXT	0～65535 字节	长文本
MEDIUMBLOB	0～16777215 字节	中等长度二进制数据
MEDIUMTEXT	0～16777215 字节	中等长度长文本
LONGBLOB	0～4294967295 字节	极大长度二进制数据
LONGTEXT	0～4294967295 字节	极大长度文本数据

3．字段中的字节大小和字符大小问题

例如 CHAR(255)，CHAR()括号内是最大字符数，因此不管何种字符集，该字段的长度都是 255 字符，对应的字节数根据字符集不同而不同。而 VARCHAR()括号内是最大字节数，任何字符编码下的最大字节数是相同的，比如 MySQL 中 UTF8 字符集每个字符占用 3 字节，UTF8MB4 字符集每个字符占用 4 字节，因此 UTF8 字符集下 VARCHAR 的最大字符数为 65535/3 = 21845，UTF8MB4 字符集下 VARCHAR 的最大字符数为 65535/4 = 16384。

只有文本类型的数据存在字符集选项，二进制数据类型没有字符集选项，排序和比较时基于列字节的数字值。

7.9　数据表操作

7.9.1　创建数据表

MySQL 中创建数据表需要提供以下信息：

- 数据表名称。
- 数据表字段名称和字段属性。
- 数据表选项。

创建数据表的语法如下：

```
CREATE TABLE table_name(
 column_name column_attribute
```

```
) table_options;
```

仅在数据表不存在时创建数据表的语法如下：

```
CREATE TABLE IF NOT EXISTS table_name(
  column_name column_attribute
) table_options;
```

下面是创建用户表的示例：

```
CREATE TABLE `user` (
 `id` int(10) unsigned NOT NULL AUTO_INCREMENT,
 `username` varchar(20) NOT NULL,
 `password` char(32) NOT NULL,
 `sex` tinyint(3) unsigned NOT NULL DEFAULT '1',
 `created_at` datetime NOT NULL,
 PRIMARY KEY (`id`),
 KEY `uq_username` (`username`)
) ENGINE=InnoDB DEFAULT CHARSET=utf8;
```

- unsigned：只能用来修饰数值类型，表示无符号数值。
- NOT NULL：可以修饰任何字段类型，表示该字段不允许 NULL 值。
- AUTO_INCREMENT：只能修饰整数值，表示自增，一般用于主键，插入数据时会自动加 1。
- DEFAULT：用来指定默认值。
- PRIMARY KEY：用来指定主键。
- KEY：用来创建自定义索引。
- ENGINE=InnoDB：用来指定存储引擎，这是 MySQL 默认的存储引擎，建议大家使用。
- DEFAULT CHARSET=utf8：用来设置默认字符集，该表中的文本型字段默认使用该字符集。

"`" 是反引号（键盘 Tab 键上面的那个键），不要和单引号 "'" 混淆。如果使用的字段名是 MySQL 关键字，则需要使用反引号将该字段包起来，否则 SQL 语句将报错。

7.9.2 删除数据表

MySQL 中删除数据表只需要一条命令，删除之后数据表消失，因此在删除数据表时必须格外小心。

```
DROP TABLE table_name;
```

7.9.3 添加字段

MySQL 支持在已存在的表上动态添加字段，添加字段的过程中该表将被锁住。
MySQL 添加字段的语法如下：

```
ALTER TABLE table_name ADD COLUMN column_name column_attribute [FIRST | AFTER
    column_name];
```

下面的示例是在 user 表中添加 updated_at 字段。

```
ALERT TABLE `user` ADD COLUMN updated_at TIMESTAMP NOT NULL ON UPDATE
    CURRENT_TIMESTAMP;
```

下面的示例是在 user 表的最前面添加 user_id 字段。

```
ALTER TABLE `user` ADD COLUMN user_id int unsigned NOT NULL FIRST;
```

以下的示例是在 user 表的 created_at 字段后添加 updated_at 字段。

```
ALTER TABLE `user` ADD COLUMN updated_at TIMESTAMP NOT NULL ON UPDATE
    CURRENT_TIMESTAMP AFTER created_at;
```

7.9.4　删除字段

MySQL 删除字段的语法如下:

```
ALTER TABLE table_name DROP COLUMN column_name;
```

下面的示例是删除 user 表中的 updated_at 字段。

```
ALTER TABLE `user` DROP COLUMN `updated_at`;
```

7.9.5　修改字段

对于已存在的字段，MySQL 也可以修改其定义。

MySQL 修改已存在字段的语法如下（column_attribute 为完整的字段属性）:

```
ALTER TABLE table_name MODIFY COLUMN column_name column_attribute;
```

下面的示例是修改 user 表中的 created_at 字段。

```
ALTER TABLE `user` MODIFY COLUMN `created_at` timestamp NOT NULL ON UPDATE
    CURRENT_TIMESTAMP(0);
```

7.10　数据操作

7.10.1　插入数据

MySQL 使用 INSERT INTO 来插入数据，语法如下:

```
INSERT INTO table_name (field1, field2, ... fieldN) VALUES (value1, value2, ...
    valueN), (value1, value2, ... valueN)
```

如果数据是字符串类型，则需要使用单引号或双引号包裹字符串数据。

下面的示例是向 user 表中插入数据。

```
INSERT INTO `user` (username, password, created_at) VALUES ('xialei', 'secret',
    '2019-11-10 23:00:00');
```

添加数据的时候还可以不指定字段名进行插入，此时 value 需要指定所有字段，但是不建议这样做，一旦数据表结构发生更改，应用程序可能发生错误。

假设 user 表中的字段为 id、username、password、created_at，下面是不指定字段名插入数据的示例。

```
INSERT INTO `user` VALUES (0, 'xialei', 'secret', '2019-11-10 23:00:00');
```

下面是批量插入用户表的示例。

```
INSERT INTO `user` (username, password, created_at) VALUES ('xialei', 'secret',
            '2019-11-10 23:00:00'), ('xialei1','secret1','2019-11-10
            23:00:00');
```

7.10.2 查询数据

MySQL 使用 SELECT 来查询数据，查询语法如下：

```
SELECT column1, column2, ... columnN FROM table_name [WHERE] [LIMIT] [ORDER BY]
```

- 查询语句中可以查询一个或多个表。
- 指定字段名返回指定字段，也可以使用*来查询所有字段。
- 指定 WHERE 来筛选符合条件的数据。
- 指定 LIMIT 来返回指定条数的数据，默认返回所有。
- 指定 ORDER BY 来对结果进行排序。

下面的示例是查询 user 表中所有的数据。

```
SELECT * FROM `user`;
```

下面的示例是查询 user 表中所有的 username 字段。

```
SELECT username FROM `user`;
```

下面的示例是查询 user 表中满足 sex 为 1 且 age>30，按照 ID 降序排序的最前面的 10 条记录。

```
SELECT * FROM `user` WHERE sex=1 AND age>30 ORDER BY `id` DESC LIMIT 10;
```

MySQL 支持强大的查询语法，常用的查询操作符如下。

- field=value：查询 field 的值为 value 的记录。
- field>value：查询 field 的值大于 value 的记录。
- field>=value：查询 field 的值大于等于 value 的记录。

- field!=value：查询 field 的值不等于 value 的记录。
- field<value：查询 field 的值小于 value 的记录。
- field<=value：查询 field 的值小于等于 value 的记录。
- field is null：查询 field 的值为 null 的记录。
- field is not null：查询 field 的值不为 null 的记录。
- field in (value1, value2, ..., valueN)：查询 field 的值在列表的记录。
- field like '%value%'：查询 field 的值包含 value 值。

如果 value 是字符串，传入 SQL 时需要使用单引号包裹字符串。

多个查询条件中可以使用 AND 或 OR 来组合。

下面的示例是查询 user 表中"username 为 xialei 且 status 为 1"或者"username 为 demo 且 status 为 0"的记录。

```
SELECT * FROM `user` WHERE (username='xialei' AND status = 1) OR (username = 'demo'
                    AND status = 0);
```

7.10.3 修改数据

MySQL 使用 UPDATE 来修改数据，语法如下：

```
UPDATE table_name SET field1=value1,field2=value2,fieldN=valueN [WHERE] [LIMIT]
```

- 可以更新一个或几个字段。
- 可以更新一条或几条甚至所有数据（取决于 LIMIT）。默认情况下没有 LIMIT，就表明要更新所有数据。

WHERE 子句和 LIMIT 子句的语法和查询数据中语法一致。

下面的示例是找到 user 表中 username 字段为 demo 的记录，把这些记录的 status 字段更新为 1。

```
UPDATE `user` SET status=1 WHERE username ='demo';
```

如果满足条件的数据很多，可以指定要更新的条数，可以指定 LIMIT 参数。

下面的示例是更新 user 表中 status 为 1 的前 10 条记录。

```
UPDATE `user` SET status=0 WHERE status=1 LIMIT 10;
```

7.10.4 删除数据

MySQL 使用 DELETE FROM 来删除数据，语法如下：

```
DELETE FROM table_name [WHERE] [LIMIT]
```

- 如果没有指定 WHERE 子句或 LIMIT 子句，表中的所有数据将被删除。
- WHERE 子句和 LIMIT 子句的语法和查询数据中语法一致。

下面的示例是删除 user 表中 status 为 0 的记录。

```
DELETE FROM `user` WHERE status=0;
```

下面的示例是删除 user 表中 status 为 0 的前 10 条记录。

```
DELETE FROM `user` WHERE status=0 LIMIT 10;
```

7.11　本章小结

本章我们学习了 MySQL 数据库的基础知识和一些基本 SQL 语句的操作。实际开发中直接操作 SQL 会降低我们的开发效率，此外还会带来潜在的 Bug，因此在实际开发中一般会使用 ORM 框架。

回顾一下本章所学：

- MySQL 服务器的安装。
- 数据库索引的概念。
- 数据库事务的概念与使用。
- 关联关系的概念。
- MySQL 的基本操作。

MySQL 也有优秀的 ORM 框架，接下来的一章中我们将学习 MySQL 的 ORM 框架 Sequelize。

第 8 章

ORM 框架——Sequelize

本章内容

- Sequelize 模型定义和扩展
- Sequelize 关联关系
- Sequelize 对数据的操作
- Sequelize 事务的使用

8.1 ORM

ORM（Object Relational Mapping，对象关系映射），通过实体对象的语法，完成对关系数据库的操作。如图 8-1 所示。

图 8-1

ORM 把数据库中的术语映射为面向对象编程中的概念。

- 数据表→类
- 数据行→类实例
- 数据列→实例属性

ORM 使用面向对象的操作，屏蔽了数据库的访问细节，因此几乎可以不编写 SQL 语句。当然，即使使用 ORM 技术，对于关系数据库的知识以及 SQL 语言还是需要了解的。

8.2 Sequelize 简介

Sequelize 是 Node.js 操作关系数据库的一个 npm 模块，基于 Promise 开发，包含很多特性：数据库模型映射、事务处理、模型属性校验、关联映射、自动创建数据表等等，支持 MySQL、PostgreSQL、MS SQL 在内的主流 SQL 数据库。

Sequelize 采用了抽象设计，并不绑定任何具体的数据库实现，因此我们除了需要安装 sequelize 模块之外，还需要安装 mysql2 模块。

mysql2 可以说是 mysql 的升级版本，原生支持 Promise。

```
npm install sequelize mysql2 -save
```

8.3 快速开始

Sequelize 一个强大的特性是自动建表，因此我们在定义数据对象模型之后通过 Sequelize 即可在数据库自动建表。

下面是一个 MySQL 数据库创建用户的示例。

```js
// 配置 sequelize
const { Sequelize, Model, DataTypes } = require('sequelize');
const sequelize = new Sequelize({
    dialect: 'mysql',
    host: 'localhost',
    port: 3306,
    username: 'root',
    password: 'root',
    database: 'test',
    timezone: '+08:00',
    pool: {
        max: 10,
        min: 0,
    },
    logging: false
});

// 定义模型

class User extends Model { }
```

```js
User.init({
    id: { type: DataTypes.INTEGER({unsigned: true}), primaryKey: true,
    autoIncrement: true, comment: '用户 ID' },
    username: { type: DataTypes.STRING(40), unique: true, allowNull: false, comment:
'账号' },
    password: { type: DataTypes.CHAR(64), allowNull: false, comment: '密码' }
}, { sequelize: sequelize, tableName: 'user', modelName: 'user' });

// 同步数据库

sequelize.sync()
    .then(() => {
        return User.create({
            username: 'xialei',
            password: 'password'
        });
    })
    .then((user) => {
        console.log(user.toJSON());
    })
    .catch((err) => {
        console.error(err);
    });
```

首先我们配置了数据库连接，接下来定义了一个 User 模型，并进行字段和模型选项设置，然后使用 sequelize.sync() 将模型同步到数据库（自动建表），最后插入了一条用户数据。

执行该文件将会在数据库中自动创建数据表 user，建表 SQL 如下：

```sql
CREATE TABLE `user` (
  `id` int(10) unsigned NOT NULL AUTO_INCREMENT COMMENT '用户 ID',
  `username` varchar(40) NOT NULL COMMENT '账号',
  `password` char(64) NOT NULL COMMENT '密码',
  `createdAt` datetime NOT NULL,
  `updatedAt` datetime NOT NULL,
  PRIMARY KEY (`id`),
  UNIQUE KEY `username` (`username`)
) ENGINE=InnoDB DEFAULT CHARSET=utf8mb4;
```

createdAt 和 updatedAt 字段是 Sequelize 自动添加的，一般情况下我们不需要手动维护这两个字段。

8.4 构造方法

Sequelize 构造方法需要传入一个初始化配置，常用的配置选项如下：

- dialect：底层数据库。支持 MySQL、PostgreSQL、MS SQL、MariaDB（MySQL 的一个分支）和 Sqlite。
- dialectModule：自定义底层数据库驱动。
- storage sqlite：数据库存储位置。默认值为:memory:，存储在内存。
- database：数据库名称。
- username：数据库账号。
- password：数据库密码。
- host：数据库连接地址。默认值为 localhost。
- port：数据库连接端口。
- ssl：是否启用 SSL 连接。默认值为 false。
- timezone：时区。默认+00:00。
- omitNull：是否将 null 值提交到数据库。默认值为 false。
- pool：连接池选项。
  - max：连接池最大连接数。默认值为 5。
  - min：连接池最小连接数。默认值为 0。
  - idle：连接空闲时间，空闲超过该时间将被释放，单位毫秒。
  - acquire：重试获取连接时的超时时间，单位毫秒，超过该时间将抛出错误。
  - evict：空闲扫描间隔，单位毫秒，Sequelize 每隔一段时间将扫描并回收空闲连接。
- logging：是否显示 SQL 日志。默认值为 true。
- benchmark：是否显示 SQL 执行时长。默认值为 false。

指定 dialect 时 Sequelize 会自动检测底层模块是否安装，如果未安装底层模块，Sequelize 会提示错误，大家根据提示错误来安装对应模块即可。

下面我们来看一个 sqlite3 示例，数据存储在内存中。

使用 sqlite3 之前需要安装 sqlite3 模块：

```
npm install sqlite3 -save
```

代码部分我们只需要变更构造 Sequelize 的参数，其他代码不用变更：

```
const sequelize = new Sequelize({
    dialect: 'sqlite',
    storage: ':memory:',
    logging: true,
    benchmark: true
});
```

8.5 数据类型

Sequelize 支持丰富的数据类型，有些是所有数据库都支持的，有些是部分数据库支持的，建议尽量使用都支持的数据类型，这样方便在以后进行数据库间的切换。

表 8-1 展示了 Sequelize 中 MySQL 驱动常用的数据类型。

表 8-1　Sequelize 中 MySQL 驱动常用的数据类型

Sequelize 类型	MySQL 类型	说明
TINYINT	TINYINT	小整数值
INTEGER	INT	大整数值
BIGINT	BIGINT	极大整数值
DECIMAL	DECIMAL	实数
STRING	VARCHAR	变长字符串
CHAR	CHAR	定长字符串
TEXT	TEXT	文本
BOOLEAN	TINYINT	布尔类型
DATE	DATETIME	日期时间
DATEONLY	DATE	日期
TIME	TIME	时间

下面的示例是创建无符号整型:

```
DataTypes.INTEGER({unsigned: true})
```

整型 TINYINT、INTEGER、BIGINT 支持的选项如下:

- length: 显示宽度。比如 INT(2),不管数值是多少,最少可以看到 2 位数,不足补零,超过 2 位数显示原始值。显示宽度只影响显示,对原始数据的存储没有影响。
- unsigned: 是否为无符号数。默认为 false。
- zerofill: 是否进行零填充。

下面的示例是创建长度为 20 的变长字符串。

```
DataTypes.STRING({length: 20, binary: false});
DataTypes.STRING(20);
```

STRING、CHAR、TEXT 类型都支持 length 选项,STRING、CHAR 还额外支持 binary 选项。

- length: 字符串的长度。
- binary: 是否使用二进制存储。默认值为 false。

一般情况下很少会使用二进制字符串,因此我们不需要传递 binary 选项。

日期时间类的字段不需要传递选项,直接使用指定类型创建即可。

8.6 模型定义

模型可以说是 Sequelize 的核心，也是内容比较多的部分，下面我们一起来学习一下 Sequelize 中模型的定义。

定义模型需要两步：

（1）继承 Sequelize 内置的 Model。
（2）使用 init 方法初始化模型字段和模型选项。

```
class User extends Model {}
User.init({
  id: { type: DataTypes.INTEGER({ unsigned: true }), primaryKey: true,
    autoIncrement: true, comment: '用户ID' },
  username: { type: DataTypes.STRING(40), unique: true, allowNull: false, comment:
    '账号' },
  password: { type: DataTypes.DATEONLY, allowNull: false, comment: '密码' }
}, { sequelize: sequelize, tableName: 'user', modelName: 'user' });
```

init 方法第 1 个参数为字段设置，第 2 个参数为模型选项。

我们先来看看字段设置。

8.6.1 字段设置

字段设置是一个对象，对象的键对应数据库的字段名，对象的值是该字段的数据类型和其他属性配置。

字段设置支持以下两种方式：

- 直接使用数据类型进行设置。
- 使用完整的字段选项。

先来看看直接使用数据类型进行设置的例子。

```
User.init({
  username: DataTypes.STRING(40)
})
```

生成数据字段的 SQL 语句如下：

```
`username` varchar(40) DEFAULT NULL
```

由此可知直接使用数据类型进行设置时，默认可以存储 NULL 值。

再来看看使用完整的字段选项进行设置的例子。

```
User.init({
```

```
    id: { type: DataTypes.INTEGER({ unsigned: true }), primaryKey: true,
    autoIncrement: true }
});
```

生成数据字段的 SQL 语句如下：

```
`id` int(10) unsigned NOT NULL AUTO_INCREMENT,
PRIMARY KEY (`id`)
```

常用的字段选项如下：

- type：数据类型。
- unique：是否唯一索引。
- primaryKey：是否主键。
- autoIncrement：是否自增。
- comment：字段注释。

8.6.2 模型选项

模型选项用来定义模型的行为属性，Sequelize 的模型选项非常丰富，我们先来看看总体的选项，然后再一一学习。

常用的模型选项如下：

- omitNull：是否提交 null 值到数据库。默认为 false。
- timestamps：是否自动添加创建时间和修改时间字段。默认为 true。
- paranoid：是否启用软删除，启用软删除的情况下将不会真正删除数据，而是设置 deletedAt 字段的值，当 deletedAt 字段有值时逻辑上就是已删除的数据。默认为 false。
- underscored：是否将驼峰命名的 JavaScript 键映射到数据库字段时转化为下划线，例如 JavaScript 中定义属性为 createdAt，在数据库建表时会创建 created_at 字段。默认 false。
- modelName：模型名称。默认值为类名。
- createdAt：设置创建时间字段的字段名，该选项只有在 timestamps 为 true 时工作。
- updatedAt：设置更新时间字段的字段名，该选项只有在 timestamps 为 true 时工作。
- deletedAt：设置删除时间字段的字段名，该选项只有在 timestamps 为 true 时工作。
- tableName：表名称，默认使用类名。
- engine：存储引擎，默认 InnoDB。
- charset：字符集。
- comment：字段注释。
- collate：字符集校对规则。
- initialAutoIncrement：自增主键的初始值。
- hooks：数据库钩子，比如可以在创建记录前添加默认值、格式化等操作。
- validate：模型验证规则定义。

hooks 和 validate 的内容比较多，接下来我们单独对这两方面的内容进行学习。

8.6.3 Hooks

Sequelize 数据库钩子函数也叫作生命周期函数，在 Sequelize 运行过程中会被调用来完成预定的功能。比如数据验证、数据处理等等业务无关的操作可以在生命周期函数中处理。

生命周期顺序

了解 Sequelize 的生命周期顺序是非常重要的，这样才可以明白执行流程，让代码按照预期运行，否则可能会发生值被覆盖或者其他情况。

全部的生命周期顺序如下：

```
beforeBulkCreate(instances, options)        // 批量创建数据之前
beforeBulkDestroy(options)                  // 批量删除数据之前
beforeBulkUpdate(options)                   // 批量更新数据之前
beforeValidate(instance, options)           // 验证之前
validate // 验证过程
afterValidate(instance, options)            // 验证完毕
 - or -
validationFailed(instance, options, error) // 或者验证失败，验证失败不会进入下面的生命
    周期

beforeCreate(instance, options)             // 创建数据之前
beforeDestroy(instance, options)            // 删除数据之前
beforeUpdate(instance, options)             // 更新数据之前
beforeSave(instance, options)               // 保存数据之前
beforeUpsert(values, options)               // 替换数据之前
create         // 创建数据
destroy        // 删除数据
update         // 更新数据
afterCreate(instance, options)              // 创建数据之后
afterDestroy(instance, options)             // 删除数据之后
afterUpdate(instance, options)              // 更新数据之后
afterSave(instance, options)                // 保存数据之后
afterUpsert(created, options)               // 替换数据之后
afterBulkCreate(instances, options)         // 批量创建之后
afterBulkDestroy(options)                   // 批量删除之后
afterBulkUpdate(options)                    // 批量更新之后
```

下面我们以注册账号为例来说明一下执行经过的生命周期。

```
beforeValidate   // 验证之前，此时要设置验证规则
validate         // 调用验证函数进行验证
afterValidate(instance, options)            // 验证通过
 - or -
validationFailed(instance, options, error) // 验证失败，注册流程结束
beforeCreate(instance, options)             // 创建账号之前
beforeSave(instance, options)               // 保存数据之前
create // 写入数据库
afterCreate(instance, options)              // 创建账号之后
```

```
afterSave(instance, options)        // 保存数据之后
```

beforeSave 在创建和编辑操作时都会回调，例如对于密码的加密操作可以在 beforeSave 进行处理。

下面来看一个用户系统中对密码加密的示例。

```
// 定义模型和生命周期函数
User.init({
    id: { type: DataTypes.INTEGER({ unsigned: true }), primaryKey: true,
    autoIncrement: true, comment: '用户 ID' },
    username: DataTypes.STRING(40),
    password: { type: DataTypes.CHAR(64), allowNull: false, comment: '密码' }
}, {
    sequelize: sequelize, tableName: 'user', modelName: 'user', paranoid: true,
    hooks: {
        beforeSave(user) { // beforeSave 钩子
            if (user.changed('password')) {
                user.password =
crypto.createHash('sha256').update(user.password).digest('hex');
            }
        }
    }
});
// 修改用户 ID 为 1 的密码
User.findByPk(1).then((user) => {
    user.password = 'password';
    return user.save();
}).then((user) => {
    console.log(user.toJSON());
});
```

在 beforeSave 函数中我们使用 user.changed('password') 来判断当前 user 的 password 是否有更改，如果产生更改，则需要对新密码进行加密。

合理的使用生命周期函数能够将逻辑分离，使我们能更加专心地开发业务逻辑。

8.6.4 生命周期函数

Sequelize 中有三种方式定义生命周期函数，推荐使用模型选项和模型类静态方法进行定义。

1. 模型选项方式

```
User.init(properties, {
  hooks: {
    beforeSave(user) { // 生命周期函数

    }
  }
})
```

2. 模型类静态方法

```
User.beforeSave((user) => {
  if (user.changed('password')) {
      user.password =
    crypto.createHash('sha256').update(user.password).digest('hex');
   }
});
```

3. addHook 方法（不推荐）

```
User.addHook('beforeSave', (user, options) => {
 if (user.changed('password')) {
      user.password =
    crypto.createHash('sha256').update(user.password).digest('hex');
   }
});
```

addHook 方法中接收的生命周期名称是字符串，如果编码时输入错误，会因为无法触发生命周期函数导致潜在的问题，而且比较难以排查，因此不推荐使用。

8.6.5 模型验证器

通过模型验证器，可以验证模型的每个字段是否符合设置的规则，如果不符合规则，就不允许写入数据库。

模型验证器的运行是 Sequelize 自动处理的，当然也可以手动调用 validate 方法进行验证。

1. 定义验证规则

在定义模型字段的时候，可以传递 validate 配置来定义验证规则，下面是一个设置邮箱字段的例子。

```
User.init({
 email:{
   type: DataTypes.STRING(40),
   validate:{
     isEmail: true
   },
   allowNull: false
 }
})
```

该模型的 email 字段必须不为空且是标准的邮箱格式，否则将无法存入数据库。

2. 内置验证规则

Sequelize 内置了非常丰富的验证规则，下面我们一起来看看。

```
validate:{
   is: ["^[a-z]+$",'i'],        // 正则表达式，第1个元素为表达式内容，第2个元素为表达式标记，
                                // 如 i、g 等
   is: /^[a-z]+$/i,             // 标准的正则表达式语法
```

```
    not: ["[a-z]",'i'],          // 同 is，不过只允许不匹配的字符串
    isEmail: true,               // 邮箱
    isUrl: true,                 // 超链接
    isIP: true,                  // IPV4 或 IPV6 地址
    isIPv4: true,                // IPV4 地址
    isIPv6: true,                // IPV6 地址
    isAlpha: true,               // 26 个字母
    isAlphanumeric: true,        // 26 个字母+10 个数字
    isNumeric: true,             // 数字
    isInt: true,                 // 整数
    isFloat: true,               // 小数
    isLowercase: true,           // 小写字母
    isUppercase: true,           // 大写字母
    notNull: true,               // 不能为 null
    isNull: true,                // 只能为 null
    notEmpty: true,              // 不能为空字符串
    equals: 'specific value',    // 只能是指定值
    contains: 'foo',             // 必须包含指定的子串
    notIn: [['foo', 'bar']],     // 不能在给定的数组中
    isIn: [['foo', 'bar']],      // 只能在给定的数组中
    notContains: 'bar',          // 不能包含给定的子串
    len: [2,10],                 // 只能包含 2-
    isDate: true,                // 日期
    isAfter: "2011-11-05",       // 只能是给定日期之后的日期
    isBefore: "2011-11-05",      // 只能是给定日期之前的日期
    max: 23,                     // 数字，最大 23
    min: 23,                     // 数字，最小 23
}
```

内置验证规则的使用非常简单，在定义模型字段的时候配置 validate 属性即可。下面是一些示例。

```
// 用户名只能是数字和字母组成，最少 6 位，最多 20 位
username: {
    type: DataTypes.STRING(40),
    validate: {
        isAlphanumeric: true,
        len: [6, 20],
    }
}

// 生日只能在 2019-10-01 之前
birthday: {
  type: DataTypes.DATEONLY,
  validate: {
     isBefore: '2019-10-01'
  }
}

// 必须是有效的 11 位手机号码
mobile: {
```

```
   type: DataTypes.CHAR(11),
   validate: {
      is: /^1[3-9]\d{9}$/
   },
   allowNull: false
}
```

3. 自定义验证

某些场景下，Sequelize 内置的验证器并不能满足我们的需求，必须需要执行一些逻辑代码才可以。

比如在教务系统中添加课程时会填写开课和结课日期，我们需要验证结课日期不能早于开课日期，这时只能通过自定义验证器来实现。

```
startDate: {
  type: DataTypes.DATEONLY,
  allowNull: false,
  validate: {
     isDate: true,
  }
},
endDate: {
  type: DataTypes.DATEONLY,
  allowNull: false,
  validate: {
     isDate: true,
     afterStartDate(value) {
        const startDateObj = new Date(this.startDate);
        const endDateObj = new Date(value);
        if (endDateObj.getTime() < startDateObj.getTime()) {
          throw new Error('结束日期不能早于开始日期');
        }
     }
  }
}
```

当我们在创建数据时，结束日期如果早于开始日期，验证将不会通过。

自定义验证函数抛出错误对象即可返回验证失败的消息。

4. 异步验证

有些场景下执行的自定义验证函数有可能是异步的，比如访问了数据库或者 HTTP 接口，幸运的是，Sequelize 的自定义验证函数是支持异步的，因此我们可以使用 async/await 来编写异步验证。

```
status: {
  allowNull: false,
  validate: {
     async remoteValidate(value) {
        const result = await httpRequest(value); // httpRequest 是返回 Promise 的接口
        if (result.errcode) {
```

```
      throw new Error('status 字段验证失败');
    }
  }
 }
}
```

5. 验证器与 allowNull

如果定义模型字段时设置了 allowNull 为 false，在使用时传递了 null 值，将跳过所有验证器（内置验证器和自定义验证函数）并抛出错误。

比如注册账号时，账号不允许为空，且只能由字母和数字组成，如果传入了 null，将直接抛出错误。

```
username: {
  type: DataTypes.STRING(20),
  allowNull: false,
  validate: {
    isAlphanumeric: true
  }
}

// 创建用户
User.create({
 username:null
}).then((user) => {
 console.log(user.toJSON())
}).catch((err) => {
 console.error(err);
})
```

如果定义模型字段时设置了 allowNull 为 true，在使用时传递了 null 值，则只跳过内置验证器，自定义验证函数仍将运行。

比如在绑定手机时，手机号码允许为空，在不为空的情况下必须满足有效的 11 位手机号码的格式。

```
mobile: {
  allowNull: true,
  validate: {
    validateMobile(value) {
      if (value === null) { // 为 null 时直接通过
        return;
      }
      if(!/^1[3-9]\d{9}$/.test(value)) {
        throw new Error('手机号码格式错误');
      }
    }
  }
}
```

6. 自定义验证消息

由于 Sequelize 是国外的开源项目,在验证失败时抛出的错误消息是英文的,因此在国内使用时更换为中文会更友好一点。

Sequelize 支持为每个验证规则定义错误消息,当验证失败时直接抛出我们自定义的错误消息。

自定义验证函数抛出错误消息的操作是我们触发的,因此本节只讨论内置验证器的错误消息。

```
username: {
  type: DataTypes.STRING(20),
  allowNull: false,
  validate: {
    isAlphanumeric: {
      msg: '账号只能由字母和数字组成',
    }
  }
},
mobile: {
  type: DataTypes.CHAR(11),
  allowNull: false,
  validate: {
    is: {
      msg: '手机号码不合法',
      args: /^1[3-9]\d{9}$/
    }
  }
}
```

Sequelize 的内置验证器很多,记忆每个验证器的 msg 和特定参数显得有点困难,其实这其中是有规律可循的。

本章的"内置验证规则"小节中列举了很多比较常用的内置验证规则,这些规则可以分成两类:

- 验证规则 1,格式为:验证规则名称: true/false,如 isEmail: true。
- 验证规则 2,格式为:验证规则名称:验证规则参数,如 isAfter: '2019-10-01'。

对于第 1 类验证规则,自定义错误消息的写法如下:

```
验证规则名称: {
  msg: '错误消息'
}
```

例如我们自定义邮箱验证错误时的消息:

```
isEmail: {
  msg: '邮箱格式错误'
}
```

对于第 2 类验证规则,自定义错误消息的写法如下:

```
验证规则名称: {
  msg: '错误消息',
  args: 验证规则参数
}
```

例如我们自定义手机号验证错误时的消息:

```
is: { // 正则表达式规则
  msg: '手机号码错误',
  args: /^1[3-9]\d{9}$/
}
```

8.6.6 模型方法

Sequelize 中采用了 ES6 的 class 语法定义模型类，我们可以直接在模型类中添加静态方法和实例方法。

下面是一个实例方法和静态方法的示例。

```
class User extends Model {
    getFullName() {
        return this.lastname + this.firstname;
        // this 代表 User 实例，也就是一行数据记录
    }
    static checkName(name) { // 静态方法，通过 User.checkName(name) 调用
      return true;
    }
}
User.init({
    id: { type: DataTypes.INTEGER({ unsigned: true }), primaryKey: true,
     autoIncrement: true, comment: '用户ID' },
    firstname: {
       type: DataTypes.STRING(10),
       allowNull: false,
       validate: {
           notEmpty: {
              msg: '名不能为空'
           }
       }
    },
    lastname: {
       type: DataTypes.STRING(10),
       allowNull: false,
       validate: {
           notEmpty: {
              msg: '姓不能为空'
           }
       }
    }
}, {
```

```
        sequelize: sequelize, tableName: 'user', modelName: 'user', paranoid: true,
         underscored: true
});

sequelize.sync({ force: true })
    .then(() => {
        return User.create({
            firstname: 'san',
            lastname: 'zhang'
        })
    })
    .then((user) => console.log(user.getFullName())) // 输出"zhangsan"
    .catch((err) => console.error(err));
```

8.6.7 索引

添加字段属性的时候可以比较方便地设置唯一索引，但是如果需要设置普通索引，就只能使用模型选项来定义了。

```
User.init({
    id: { type: DataTypes.INTEGER({ unsigned: true }), primaryKey: true,
     autoIncrement: true, comment: '用户ID' },
    username: {
        type: DataTypes.STRING(40),
        allowNull: false
    },
    status: {
        type: DataTypes.TINYINT,
        allowNull: false
    }
}, {
    sequelize: sequelize,
    tableName: 'user',
    modelName: 'user',
    paranoid: true,
    underscored: true,
    indexes: [
        {
            name: 'idx_username_status',
            fields: ['username', 'status'],
            unique: false,
        }
    ]
});
```

模型选项的 indexes 属性用来定义索引，该数组的每个元素都是一个独立的索引，常用的属性如下：

- name: 索引名称。

- fields：索引字段，如果是联合索引则添加多个字段名称。
- unique：是否唯一索引。

8.6.8 数据库同步

使用 Sequelize 开发新项目时一般不需要事先建立数据表，只需要定义模型，剩余工作 Sequelize 可以自动完成。当前支持创建和删除数据表：

```
sequelize.sync();    // 建立数据表，如果数据表存在则跳过
sequelize.sync({force: true});  // 强制建立数据表，如果数据表已存在将被覆盖
sequelize.drop();    // 删除数据表
```

由于.sync({force:true})是破坏性操作，因此 Sequelize 提供了额外的 match 属性用于附加的安全检查。match 接收一个正则表达式并对数据库名称进行匹配，如果通过验证则执行 sync 方法，否则不执行。

```
sequelize.sync({force:true,match:/_test$/});
    // 只有_test 结尾的数据库才会强制执行建表操作
```

8.7 模型使用

学习完模型定义有关的内容后，我们来一起看看模型的使用。

在本节中使用的模型定义如下：

```
class User extends Model {
}
User.init({
   id: { type: DataTypes.INTEGER({ unsigned: true }), primaryKey: true,
    autoIncrement: true, comment: '用户ID' },
   username: {
      type: DataTypes.STRING(40),
      allowNull: false,
      unique: true
   },
   password: {
      type: DataTypes.CHAR(64),
      allowNull: false
   }
}, {
   sequelize: sequelize,
   tableName: 'user',
   modelName: 'user',
   paranoid: true,
   underscored: true
});
```

8.7.1 插入数据

Sequelize 支持单条插入和批量插入数据两种方式,我们先来看看第一种方式。

1. 单条插入

直接调用模型类的 create 方法传入数据对象即可,插入成功将返回模型实例对象。

```
async function example() {
  try {
    await sequelize.sync({ force: true });
    const user = await User.create({
      username: 'xialei',
      password: '111111'
    });
    console.log('创建成功', user.toJSON());
  } catch (e) {
    console.log('创建失败', e);
  }
}

example();
```

2. 批量插入

将需要插入的数据构造为一个数组,调用模型类的 bulkCreate 方法即可,插入成功将返回模型实例数组。

批量插入时如果有一条数据未通过验证,则整个插入过程失败。

```
async function example() {
  try {
    await sequelize.sync({ force: true });
    const users = await User.bulkCreate([ // 批量插入
      { username: 'xialei', password: '111111' },
      { username: 'zhangsan', password: '111111' }
    ], {
      individualHooks: true // 对每个数据分别执行一遍生命周期函数
    });
    console.log('创建成功', users.map(user => user.toJSON()));
  } catch (e) {
    console.log('创建失败', e);
  }
}

example();
```

默认情况下,批量插入数据只会触发 bulk 相关的生命周期函数。如果需要为每个数据元素都执行独立的生命周期函数(比如 beforeSave),则需要传入 individualHooks 选项。

8.7.2 更新数据

更新数据分为两种方式，一种是直接使用模型类的 update 方法，该方法一般用来批量更新，返回更新的行数；另外一种是先查询出模型实例，然后再进行更新，是比较常用的方法。

1. 批量更新

模型类的 update 方法接收两个参数，第 1 个参数为更新数据，第 2 个参数为更新选项。常用的更新选项如下：

- where: 查询对象。满足该对象的记录才会被更新，该参数是可选的。
- individualHooks: 是否对每个数组执行单独的生命周期函数。默认为 false。
- limit: 更新数量。在满足查询条件的所有记录中，更新指定的行数。

下面示例是把用户名为 xialei 的所有记录中的密码字段都更新为 123456。

```
async function example() {
  try {
    const [count] = await User.update({ password: '123456' }, { // ES6 解构赋值语法
      where: { // 查询账号为 xialei 的记录
        username: 'xialei'
      },
      individualHooks: true
    });
    console.log('更新成功, 共更新 ' + count + '条记录');
  } catch (e) {
    console.log('更新失败', e);
  }
}
example();
```

2. 单条更新

单条更新需要查询出模型实例，然后设置更新数据，最后调用实例的 save 方法完成更新。关于查询相关的语法将在后文介绍。

```
async function example() {
  try {
    const user = await User.findOne({    // 先查询数据
      where: {
        username: 'xialei'
      }
    });
    if (!user) {
      throw new Error('用户不存在');
      return;
    }
```

```
            user.password = '111111';              // 设置更新数据
            await user.save();                     // 保存
            console.log('编辑成功', user.toJSON());
    } catch (e) {
        console.log('编辑失败', e);
    }
}
example();
```

8.7.3 删除数据

删除数据分为两种方式，一种是调用模型类的 delete 方法，该方法一般用来批量删除，返回删除的行数；另外一种是先查询出模型实例，然后调用 destroy 方法进行删除。

1. 批量删除

在软删除的情况下，将不会真正从数据库删除数据，只是把 deletedAt 设为当前时间。

```
async function example() {
    try {
        const count = await User.destroy({
            where: {
                username: 'xialei'
            }
        });
        console.log('删除成功，共删除 ' + count + ' 条记录');
    } catch (e) {
        console.log('编辑', e);
    }
}
example();
```

2. 单条删除

单条删除需要先查询出模型实例，然后调用 destroy 方法进行删除。

在软删除的情况下，将不会真正从数据库删除数据，只是把 deletedAt 设为当前时间。与批量删除不同的是，单条删除可以接收一个 force 选项，无视软删除设置直接将数据从数据库中删除。

```
async function example() {
    try {
        const user = await User.findOne({
            where: {
                username: 'xialei'
            }
        });
        if (!user) {
            throw new Error('用户不存在');
```

```
        }
        await user.destroy();
        console.log('删除成功');
    } catch (e) {
        console.log('删除失败', e);
    }
}
example();
```

3. 软删除

在模型定义了软删除的情况下，默认只会软删除数据，数据在数据库中依旧存在，但是通过 Sequelize 无法访问。如果需要强制删除数据，需要传入 force 选项。

下面是无视软删除设置删除数据的示例。

```
async function example() {
    try {
        const user = await User.findOne({
            where: {
                username: 'xialei'
            }
        });
        if (!user) {
            throw new Error('用户不存在');
        }
        await user.destroy({ force: true }); // 传入 force 选项
        console.log('删除成功');
    } catch (e) {
        console.log('删除失败', e);
    }
}
example();
```

8.7.4 查询数据

find 系列方法将从数据库中查询数据，并把查询结果返回给模型实例。本小节我们主要学习 find 系列的查询方法（即函数），高级的查询语法将在后面的章节中学习。

1. 查询单条记录

Sequelize 支持通过主键或者查询条件来获取指定记录。

调用模型类的 findByPK 方法并传入主键即可。

```
async function example() {
    const user = await User.findByPk(1);
    console.log(user ? user.toJSON() : '未查询到数据');
}
example();
```

调用模型类的 findOne 方法并传入 where 选项。

```
async function example() {
    const user = await User.findOne({
     attributes: ['name'], // 指定返回字段
     where: {
        username: 'xialei'
     }
    });
    console.log(user ? user.toJSON() : '未查询到数据');
}
```

2. 查询多条数据

findAll 用来查询满足条件的所有记录，下面是一些示例。

```
// 查询所有数据
async function example() {
    const users = await User.findAll();
    console.log(users);
}
// 查询 recommend 为 1 的 11 ~ 20 条数据，先按照 status 倒序，然后按照 ID 倒序排列
async function example() {
    const users = await User.findAll({
        where: {
            recommend: 1
        },
        order: [
            ['status', 'DESC'], // DESC 为倒序（从大到小排列），ASC 为正序（从小到大排列）
            ['id', 'DESC']
        ],
        offset: 10,
        limit: 10
    });
    console.log(users);
}
```

- order：设置排序。
- offset：设置记录集偏移，一般用来进行分页。
- limit：限制查询结果的行数。

3. 查询或创建单条数据

findOrCreate 方法用来查询单条数据，如果数据不存在则将自动创建这条数据，并返回查询或创建的模型实例。

findOrCreate 方法返回一个数组，数组的第 1 个元素为模型实例，第 2 个元素标记模型实例是通过查询还是创建得到的。

下面先查询 username 为 xialei 的记录，如果记录不存在则自动创建。

```
async function example() {
    const [user, created] = await User.findOrCreate({ // ES6 解构赋值语法
```

```
    where: {
        username: 'xialei'
    },
    defaults: { // 若未查询到数据，则作为默认值插入数据库
        username: 'xialei',
        password: '111111'
    }
});
if (created) {
    console.log('创建用户成功', user.toJSON());
} else {
    console.log('查询用户成功', user.toJSON());
}
}
example();
```

4. 同时查询数据列表和数据总数

findAndCountAll 方法用来查询满足条件的数据列表和数据总数，在分页场景下用得比较多，该方法返回一个包含 rows 和 count 属性的对象，rows 是模型实例列表，count 是数据总数。

```
async function example() {
    const { rows, count } = await User.findAndCountAll({
        where:{
            status: 1
        }
    });
    console.log(rows.map(row => row.toJSON()), count);
}
example();
```

5. 统计数据

count 方法用来统计所有数据或者满足条件的数据的总数量，在分页场景下用得比较多。

```
// 不带查询条件的统计
async function example() {
    const count = await User.count();
    console.log('用户总数为: ' + count);
}
```

```
// 统计满足条件的数据的总数量
async function example() {
    const count = await User.count({
        where: {
            status: 1
        }
    });
    console.log('status 为 1 的用户数为: ' + count);
}
```

8.7.5 查询语法

关系数据库中查询是相对比较复杂的语法，因此我们单独作为一节进行学习。

无论是进行查询、更新和删除，都需要使用查询语法来筛选符合条件的数据。

where 通常从 attribute: value 中获取一个对象，其中 value 可以是值，也可以是其他运算符的键。

1. 基本使用

我们直接看一些简单的示例，示例中会涉及运算符。

```
const Op = Sequelize.Op;
Post.findAll({
  where: {
    authorId: 2
  }
});
// SELECT * FROM post WHERE authorId = 2

Post.findAll({
  where: {
    authorId: 12,
    status: 'active'
  }
});
// SELECT * FROM post WHERE authorId = 12 AND status = 'active';
Post.findAll({
  where: {
    [Op.or]: [{authorId: 12}, {authorId: 13}]
  }
});
// SELECT * FROM post WHERE authorId = 12 OR authorId = 13;

Post.findAll({
  where: {
    authorId: {
      [Op.or]: [12, 13]
    }
  }
});
// SELECT * FROM post WHERE authorId = 12 OR authorId = 13;

Post.destroy({
  where: {
    status: 'inactive'
  }
});
// DELETE FROM post WHERE status = 'inactive';
```

```javascript
Post.update({
  updatedAt: null,
}, {
  where: {
    deletedAt: {
      [Op.ne]: null
    }
  }
});
// UPDATE post SET updatedAt = null WHERE deletedAt IS NOT NULL;
```

2. 运算符

Sequelize 内建了很多运算符，比如 Like、>、in 等，下面是常用的一些运算符。

```javascript
const Op = Sequelize.Op;
[Op.and]: [{a: 5}, {b: 6}]        // (a = 5) AND (b = 6)
[Op.or]: [{a: 5}, {a: 6}]         // (a = 5 OR a = 6)
[Op.gt]: 6,                       // > 6
[Op.gte]: 6,                      // >= 6
[Op.lt]: 10,                      // < 10
[Op.lte]: 10,                     // <= 10
[Op.ne]: 20,                      // != 20
[Op.eq]: 3,                       // = 3
[Op.is]: null                     // IS NULL
[Op.not]: true,                   // IS NOT TRUE
[Op.between]: [6, 10],            // BETWEEN 6 AND 10
[Op.notBetween]: [11, 15],        // NOT BETWEEN 11 AND 15
[Op.in]: [1, 2],                  // IN [1, 2]
[Op.notIn]: [1, 2],               // NOT IN [1, 2]
[Op.like]: '%hat',                // LIKE '%hat'
[Op.notLike]: '%hat'              // NOT LIKE '%hat'
[Op.startsWith]: 'hat'            // LIKE 'hat%'
[Op.endsWith]: 'hat'              // LIKE '%hat'
[Op.substring]: 'hat'
```

8.7.6 事务

MySQL 事务的概念在上一章中已经介绍过，下面我们来看看 Sequelize 中事务的使用。

Sequelize 需要先调用 sequelize 实例的 transaction 方法获取事务实例，将该事务实例传递到模型操作中即可。

Sequelize 事务分为自动事务和手动事务，自动事务接收一个闭包函数，函数执行过程中如果没有出错，则提交事务，闭包函数的返回值作为事务的返回值。如果函数执行过程中出错，则回滚事务并抛出错误。

1. 自动事务

调用 sequelize.transaction 方法即可开启事务，该方法还可以设置事务隔离级别。

自动事务只有在闭包函数未抛出错误时才会提交，否则将会回滚事务。

下面是一个隔离级别为串行化的事务示例。

```js
async function example() {
    const result = await sequelize.transaction({
        isolationLevel: Transaction.ISOLATION_LEVELS.SERIALIZABLE // 设置隔离级别
    }, async (transaction) => {        // 闭包函数
        const user = await User.create({
            username: 'xialei',
            password: '111111'
        }, { transaction: transaction });
        // 必须传递事务对象，否则create操作将不被事务管理

        const user2 = await User.create({
            username: 'demo',
            password: '111111'
        }, { transaction: transaction });
        return [user, user2];          // 返回数组
    });
    console.log(result.map(user => user.toJSON()));
    // result 就是[user, user2]数组，因此可以调用map
}
example();
```

2. 手动事务

手动事务也需要调用 Sequelize 的 transaction 方法，但是不能传递闭包函数，通过手动调用 transaction 实例的 commit 提交事务或者 rollback 回滚事务。

手动事务一定要调用 commit 或者 rollback，否则请求将挂起。

```js
async function example() {
    const transaction = await sequelize.transaction({
        isolationLevel: Transaction.ISOLATION_LEVELS.SERIALIZABLE
    });
    try {
        const user = await User.create({
            username: 'xialei',
            password: '111111'
        }, { transaction: transaction });
        const user2 = await User.create({
            username: 'demo',
            password: '111111'
        }, { transaction: transaction });     // 传递事务对象
        await transaction.commit();           // 提交事务
        console.log(user.toJSON(), user2.toJSON());
    } catch (e) {
        await transaction.rollback();         // 回滚事务
        throw e;        // 继续抛出错误给上层处理
    }
}
```

```
example();
```

8.8 关 联

关联关系可以说是关系数据库最重要的概念，关系数据库的由来也是因为关联关系。Sequelize 支持 hasOne、hasMany、belongsTo、belongsToMany 四种关联关系。

8.8.1 hasOne

hasOne 是普通的一对一关系，主表中的每行数据在从表中最多有一行关联数据（允许没有）。下面是模型定义，User 为主模型，Idcard 为从模型。

```
class User extends Model { }
User.init({
   name: DataTypes.STRING,
   email: DataTypes.STRING
}, {
   sequelize,
   modelName: 'user',
   underscored: true
});


class Idcard extends Model { }
User.init({
   name: DataTypes.STRING
}, {
   sequelize,
   modelName: 'idcard',
   underscored: true
});

User.hasOne(Idcard, { // 关联定义
   constraints: false
});
```

实际开发中一般只会添加外键，但是不会启用外键约束，因为使用 MySQL 的外键约束进行数据操作会引发性能问题，一般是通过应用程序来完成外键约束功能。

使用下列代码即可关闭外键约束（idcard 表的 user_id 字段依然存在）。

```
User.hasOne(Idcard, {
   constraints: false, // 关闭外键约束
});
```

如果需要自定义从表（idcard 表）中外键（user_id）的字段名称，可以传入 foreignKey 选项。

```
User.hasOne(Idcard, {
    constraints: false,
    foreignKey: 'uid', // idcard 表中的外键名称指定为 uid
});
```

1. 关联数据的插入

一对一关联支持分开创建和关联创建两种方式。分开插入也就是分别调用模型的 create 方法,而关联插入只需要调用主模型的 create 方法。

下面来看一个分开创建的示例(User 和 Idcard 复用之前的定义)。

```
async function example() {
    const uid = await sequelize.transaction(async (transaction) => {
        // 插入数据
        const user = await User.create({
            name: 'xialei',
            email: 'demo@gmail.com'
        }, { transaction: transaction });
        const idcard = await Idcard.create({
            name: 'xialei',
            userId: user.id
        }, { transaction: transaction });
        return user.id
    });
    console.log('用户创建成功, ID: ' + uid);
}

example();
```

由于需要操作两个表,因此我们启用了事务,防止出现 User 插入成功而 Idcard 插入失败的情况。

下面来看一个创建关联的示例。

```
async function example() {
    await sequelize.sync({ force: true });
    const user = await User.create({
        name: 'xialei',
        email: 'demo@gmail.com',
        idcard: { // 关联数据对象
            name: 'xialei1'
        }
    }, { include: [Idcard] }); // 需要包含关联模型,否则数据不会写入
    console.log(user.toJSON());
}

example();
```

使用关联数据插入时,需要在数据对象定义关联属性以及在插入选项中 include 关联模型。

关联插入一般用在需要同时插入新数据的场景,如果主模型中已经存在数据,那么使用第 1 种方式插入。

2. 关联查询

在查询时使用 include 包含关联模型即可。

```
async function example() {
  await sequelize.sync({force: true});
  const user = await User.findOne({
    where: {
      name: 'xialei'
    },
    include: [Idcard] // 包含关联模型
  });
  console.log(user.toJSON());
}
```

查询出来的 user 对象会包含 idcard 对象。

8.8.2 belongsTo

belongsTo 也是一对一的关联，不过此时是以从模型作为主语，从模型属于主模型。以刚才的用户和身份证为例，身份证属于用户。

```
Idcard.belongsTo(User, { // 关联定义
    constraints: false,
    foreignKey: 'uid'
});
```

1. 关联数据的插入

belongsTo 关联中分开创建数据的写法和 hasOne 一致，我们看一下关联插入。

```
async function example() {
    await sequelize.sync({ force: true });
    const idcard = await Idcard.create({
        name: 'xialei1',
        user: { // 关联数据对象
            name: 'xialei',
            email: 'demo@gmail.com'
        }
    }, {
        include: [User] // 包含关联模型
    });
    console.log(idcard.toJSON());
}
example();
```

idcard 对象会包含 user 对象。

2. 关联查询

数据查询和 hasOne 是一致的，需要使用 include 关联模型。

```
async function example() {
    const idcard = await Idcard.findOne({
        where: { uid: 1 },
        include: [User]  // 传入关联对象
    });
    console.log(idcard.toJSON());
}

example();
```

上述代码的输出如下：

```
{
  id: 1,
  name: 'xialei',
  createdAt: 2019-10-01T10:07:59.000Z,
  updatedAt: 2019-10-01T10:07:59.000Z,
  uid: 1,
  user: {
    id: 1,
    name: 'xialei',
    email: 'demo@gmail.com',
    createdAt: 2019-10-01T10:07:59.000Z,
    updatedAt: 2019-10-01T10:07:59.000Z
  }
}
```

如果需要自定义嵌套属性的名称，比如把 idcard 改成 id_card，需要进行如下操作：

- 调用 hasOne 时传入 as 选项。
- 调用查询时 include 属性设置 as 选项。

```
// 模型定义
User.hasOne(Idcard, {
    constraints: false,
    foreignKey: 'uid',
    as: 'id_card'  // 别名
});

async function example() {
    // 关联查询
    const user = await User.findOne({
        include: [{ model: Idcard, as: 'id_card' }],  // 别名
    });
    console.log(user.toJSON());
}

example();
```

上述代码的输出如下：

```
{
```

```
    id: 1,
    name: 'xialei',
    email: 'demo@gmail.com',
    createdAt: 2019-10-01T10:07:59.000Z,
    updatedAt: 2019-10-01T10:07:59.000Z,
    id_card: {
      id: 1,
      name: 'xialei',
      createdAt: 2019-10-01T10:07:59.000Z,
      updatedAt: 2019-10-01T10:07:59.000Z,
      uid: 1
    }
}
```

8.8.3 hasMany

一对多关联的语法与一对一关联的语法类似，只不过调用的是 hasMany 方法。

下面是定义 User（用户）和 Article（文章）关联的示例。

```
class User extends Model { }
User.init({
    id: { type: DataTypes.INTEGER, primaryKey: true, autoIncrement: true },
    name: DataTypes.STRING,
    email: DataTypes.STRING
}, {
    sequelize,
    modelName: 'user',
    underscored: true
});

class Article extends Model { }
Article.init({
    title: Sequelize.STRING
}, {
    sequelize,
    modelName: 'article',
    underscored: true
});

User.hasMany(Article, { // 关联定义
    constraints: false,
});
```

1. 关联数据的插入

一对多关联中支持分开插入和关联插入两种方式，与一对一的插入方式类似。

下面是分别创建一个用户和一篇文章的示例。

```
async function example() {
    await sequelize.sync({ force: true });
```

```
    const uid = await sequelize.transaction(async (transaction) => {
        const user = await User.create({            // 插入用户
            name: 'xialei',
            email: 'demo@gmail.com'
        }, { transaction: transaction });
        const article = await Article.create({      // 插入文章
            title: 'xialei',
            userId: user.id
        }, { transaction: transaction });
        return user.id
    });
    console.log('创建成功, ID: ' + uid);
}
example();
```

下面是使用关联插入新建一个用户,并插入 3 篇文章的示例。

```
async function example() {
   await sequelize.sync({ force: true });
   const user = await User.create({
       name: 'xialei',
       email: 'demo@gmail.com',
       articles: [  // 关联对象数组,每个对象将独立插入
           { title: 'Node.js 实战' },
           { title: 'Node.js 实战 2' },
           { title: 'Node.js 实战 3' },
       ]
   }, {
       include: [Article] // 包含关联模型
   });
   console.log(user.toJSON());
}
example();
```

在上述例子中,该用户有 3 篇文章,所以插入数据时需要传递数组。

2. 关联查询

hasMany 查询关联数据时使用 include 关联模型即可。

下面是查询用户和关联文章的示例。

```
async function example() {
  const user = await User.findOne({
    where: {
      name: 'xialei'
    },
    include: [Article]
  });
  console.log(user);
}
```

```
example();
```

上述代码的输出如下：

```
{
  id: 1,
  name: 'xialei',
  email: 'demo@gmail.com',
  createdAt: 2019-10-01T10:07:59.000Z,
  updatedAt: 2019-10-01T10:07:59.000Z,
  articles: [{
    id: 1,
    title: 'Node.js之路'
    createdAt: 2019-10-01T10:07:59.000Z,
    updatedAt: 2019-10-01T10:07:59.000Z,
    user_id: 1
  },
  {
    id: 2,
    title: 'Node.js之路2'
    createdAt: 2019-10-01T10:07:59.000Z,
    updatedAt: 2019-10-01T10:07:59.000Z,
    user_id: 1
  },
  {
    id: 3,
    title: 'Node.js之路3'
    createdAt: 2019-10-01T10:07:59.000Z,
    updatedAt: 2019-10-01T10:07:59.000Z,
    user_id: 1
  }
  ]
}
```

8.8.4　belongsToMany

在上一章节的学习中我们知道，多对关联关系需要中间表才可以实现。

Sequelize 中使用 belongsToMany 来定义多对多关系，下面是学生（Student）和课程（Class）的关联示例。

```
class Student extends Model { }
Student.init({
   name: DataTypes.STRING,
   email: DataTypes.STRING
}, {
   sequelize,
   modelName: 'student',
   underscored: true
});
```

```
class Class extends Model { }
Class.init({
   name: DataTypes.STRING
}, { sequelize, modelName: 'class', underscored: true })

Student.belongsToMany(Class, {
   through: 'student_class',
   constraints: false
});

Student.belongsToMany(Class, {
  through: 'student_lesson', // 中间表
  constraints: false
});
```

1. 关联数据的插入

一对多关联数据的创建与一对多关联数据的创建类似，不同的是 belongsToMany 关联的中间表是由 Sequelize 来维护的。

下面是向中间表插入学生和课程关联数据的示例。

```
async function example() {
   await sequelize.sync({ force: true });
   const student = await Student.create({
      name: 'xialei',
      email: 'demo@gmail.com',
      classes: [
         { id: 1, name: '班级1' },
         { id: 2, name: '班级2' },
      ]
   }, { include: [Class] });
   console.log(student.toJSON());
}
example();
```

上述代码执行后会自动向 student_class 中间表插入学生和课程的关联数据，不用手动维护。

2. 关联查询

belongsToMany 查询关联数据时使用 include 关联模型即可。

```
async function example() {
   await sequelize.sync();
   const students = await Student.findAll({
      include: [Class]
   });
   students.forEach((student) => {
      console.log(student.toJSON());
   });
}
```

```
example();
```

students 是一个学生列表，因此我们调用 forEach 方法进行打印，在整个查询代码中并没有出现 student_class 中间表，可以说中间过程是透明的。

上述代码输出如下：

```
{
  id: 1,
  name: 'xialei',
  email: 'demo@gmail.com',
  createdAt: 2019-11-20T03:17:06.000Z,
  updatedAt: 2019-11-20T03:17:06.000Z,
  classes: [
    {
      id: 1,
      name: '班级1',
      createdAt: 2019-11-20T03:17:06.000Z,
      updatedAt: 2019-11-20T03:17:06.000Z,
      student_class: [Object]
    },
    {
      id: 2,
      name: '班级2',
      createdAt: 2019-11-20T03:17:06.000Z,
      updatedAt: 2019-11-20T03:17:06.000Z,
      student_class: [Object]
    }
  ]
}
```

8.9　本章小结

至此，你已经学会了 Sequelize 的常用知识，并且已经知道如何结合 Web 框架来实现简单的数据库应用。然而，数据库的相关技术是比较复杂的。如果要全面了解 Sequelize 的高级功能，则需要参阅 Sequelize 的官方文档，不过在实际项目中，本章的知识点应该可以满足大部分的场景。

回顾一下本章所学：

- Sequelize 的基本使用。
- 模型的定义。
- 模型的使用。
- 关联关系。

在下一章的内容中，我们将使用目前学习过的知识开发一个微博系统。

第 9 章

微博项目开发

本章内容

- 基于 koa+sequelize 构建完整的 Web 应用

学习完 Web 框架和 MySQL 数据库相关知识后，本章我们开发一个基于 MySQL 的微博系统。

本项目使用到的依赖：

- koa
- koa-bodyparser
- koa-ejs
- koa-router
- koa-static
- mysql2
- sequelize
- nodemon

9.1 功能分析

大家几乎都使用过微博，对于微博的几大核心功能应该说比较了解，大致有以下功能

- 用户登录/注册。
- 发布/编辑/删除微博。
- 查看他人微博。

- 评论微博。
- 转发微博。
- 点赞微博。
- 关注/取消关注。

当然新浪微博的功能实现复杂很多，我们这个例子主要实现数据的展示，不会实现新浪微博那种复杂的 Feed 流功能。

本项目实现的功能如下：

- 用户登录/注册/退出登录。
- 发表/编辑/删除微博。
- 评论/回复评论/删除评论。
- 用户个人中心/修改密码。
- 用户个人首页。
- 广场（首页）。

9.2 数据模型

通过功能分析可知，用户注册需要用户模型，微博相关功能需要微博模型，而评论功能则需要评论模型。下面是微博系统所涉及的模型：

- 用户。
- 微博。
- 评论。

除了模型分析之外，还有一些场景下的字段设置，我们下面一起来看一下。

微博有自己发的和转发的，因此在微博模型中我们需要一个类型字段来标记是自己发的还是转发的。

本着数据至上的原则，在本项目中我们所有数据都采用软删除的形式来实现。

9.3 开始编码

9.3.1 初始化项目

本项目使用 koa+sequelize 进行开发，使用以下命令初始化项目：

```
npm init -y
npm install koa koa-router koa-ejs sequelize mysql2 -save
```

9.3.2 项目目录

根据职责进行目录划分，本项目的目录结构如下：

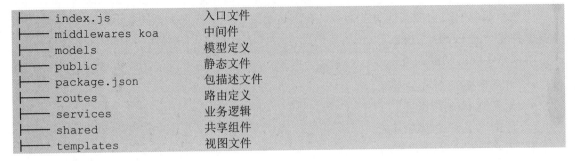

9.3.3 路由设计

根据我们需要实现的功能，路由设计如下：

- /user：用户模块。
  - /login：登录。
  - /register：注册。
  - /logout：退出登录。
  - /home：个人中心。
  - /homepage：个人主页。
  - /profile：修改资料。
- /comment：评论模块。
  - /delete：删除评论。
- / 广场：首页。
- /weibo：微博模块。
  - /publish：发布微博。
  - /edit：编辑微博。
  - /delete：删除微博。
  - /show：微博详情。
  - /comment：评论微博。
  - /share：转发微博。

9.3.4 共享组件

共享组件在整个系统中只有一个实例，适合分离到单独的目录中，项目使用的共享组件如下：

- shared/security.js：安全相关，比如哈希函数。

- shared/sequelize.js Sequelize: 实例。

```js
// shared/security.js
const crypto = require('crypto');
// SHA256 加密
exports.sha256 = function (data) {
    return crypto.createHash('sha256').update(data).digest('hex');
};
```

```js
// shared/sequelize.js
const {Sequelize} = require('sequelize');

module.exports = new Sequelize({
    dialect: 'mysql',
    host: 'localhost',
    port: 3306,
    username: 'root',
    password: 'root',
    database: 'weibo'
});
```

9.3.5 中间件

中间件是业务无关的代码，根据需要在全局导入或者在路由导入。

接下来我们看看本项目中定义的中间件。

1. 鉴权

读取 cookie 中的 userId 字段，并挂载到 ctx.state.userId 上。未登录用户的 userId 设置为 0，不会强制重定向到登录页，保证一些允许登录用户或未登录用户业务的正常使用。

```js
// middlewares/authenticate.js
module.exports = async function (ctx, next) {
    ctx.state.userId = Number(ctx.cookies.get('userId', {signed: true}));
    await next();
};
```

2. 错误处理

系统发生错误时，渲染错误页面。

本项目是基于页面的 Web 项目，因此错误提示需要渲染错误页面。

```js
// middlewares/errorHandler.js
module.exports = async function (ctx, next) {
    try {
        await next();
    } catch (e) {
        await ctx.render('error', {
            error: e.message,
            title: '错误'
```

```
    });
  }
};
```

3. 登录限制

检测 ctx.state.userId，如果为 0 则强制重定向到登录页，以保证需要登录来确认访问的页面权限。

```
// middlewares/guard.js
module.exports = async function (ctx, next) {
    if (!ctx.state.userId) {
        await ctx.redirect('/user/login');
        return;
    }
    await next();
};
```

9.3.6 模型代码

项目使用的模型文件如下：

- models/comment.js：评论。
- models/follow.js：关注。
- models/praise.js：点赞。
- models/user.js：用户。
- models/weibo.js：微博。

需要注意的是，本项目将模型定义文件拆分到 models 目录了，因此我们在导出模型的时候需要使用下面的语法：

```
const {Model} = require('sequelize');
module.exports = (sequelize, DataTypes) => {
  class 模型类 extends Model {}
  模型类.init(); // 模型设置
  // 生命周期函数
  // 关联定义
  return 模型类;
};
```

编写模型时先定义模型自己的属性，编写完所有模型之后，再来添加关联关系和生命周期函数。

下面是本项目中模型的代码：

```
// 评论 models/comment.js
const {Model} = require('sequelize');

module.exports = (sequelize, DataTypes) => {
    const Weibo = sequelize.import('./weibo');
    const User = sequelize.import('./user');
```

```js
    class Comment extends Model {
    }

    // 模型定义
    Comment.init({
        content: {type: DataTypes.STRING(140), allowNull: false, comment: '评论内容'}
    }, {
        sequelize: sequelize,
        tableName: 'comment',
        underscored: true,
        paranoid: true
    });
    // 关联定义
    Comment.belongsTo(Weibo, {    // 评论属于微博博主
        constraints: false,
        foreignKey: 'weiboId'
    });
    Comment.belongsTo(User, {     // 评论属于用户
        constraints: false,
        foreignKey: 'userId',
        as: 'user'
    });

    Comment.afterSave(async (comment) => {
        // 微博评论数+1
        await Weibo.increment({commentCount: 1}, {where: {id: comment.weiboId}});
    });
    Comment.afterDestroy(async (comment) => {
        // 微博评论数-1
        await Weibo.increment({commentCount: -1}, {where: {id: comment.weiboId}});
    });
    return Comment;
};
```

```js
// 用户 models/user.js
const {Model} = require('sequelize');
const security = require('../shared/security');

module.exports = (sequelize, DataTypes) => {
    class User extends Model {
        // 检测密码
        checkPassword(rawPassword) {
            return security.sha256(rawPassword) === this.password;
        }
    }

    User.init({
        username: {
            type: DataTypes.STRING(20),
```

```
            allowNull: false,
            validate: {
                notEmpty: {
                    msg: '账号不能为空'
                },
                len: {
                    msg: '账号长度为6-20位',
                    args: [6, 20]
                },
                isAlphanumeric: {
                    msg: '账号只能包含字母和数字'
                }
            },
            comment: '账号'
        },
        password: {type: DataTypes.CHAR(64), allowNull: false, comment: '密码'},
        nickname: {type: DataTypes.STRING(20), allowNull: false, defaultValue: '',
    comment: '昵称'},
        weiboCount: {type: DataTypes.INTEGER, allowNull: false, defaultValue: 0,
    comment: '微博数'}
    }, {
        sequelize: sequelize,
        tableName: 'user',
        underscored: true,
        paranoid: true,
        indexes: [
            {
                name: 'idx_username',
                fields: ['username']
            }
        ]
    });

    User.beforeSave((user) => {
        // 密码处理
        if (user.changed('password') && user.password.length > 0) {
            user.password = security.sha256(user.password);
        }
    });
    return User;
};


// 微博 models/weibo.js
const {Model} = require('sequelize');

module.exports = (sequelize, DataTypes) => {
    const User = sequelize.import('./user');

    class Weibo extends Model {
    }
```

```javascript
// 关联定义
Weibo.init({
    type: {type: DataTypes.TINYINT, allowNull: false, comment: '发布类型'},
    content: {type: DataTypes.STRING(140), allowNull: false, comment: '微博内容'},
    shareContent: {type: DataTypes.STRING(140), comment: '转发语'},
    praiseCount: {type: DataTypes.INTEGER, allowNull: false, defaultValue: 0, comment: '点赞数'},
    commentCount: {type: DataTypes.INTEGER, allowNull: false, defaultValue: 0, comment: '评论数'},
    shareCount: {type: DataTypes.INTEGER, allowNull: false, defaultValue: 0, comment: '转发数'}
}, {
    sequelize: sequelize,
    tableName: 'weibo',
    underscored: true,
    paranoid: true
});

Weibo.belongsTo(User, {
    constraints: false,
    foreignKey: 'userId',
    as: 'user'
});

Weibo.afterCreate(async (weibo) => {
    // 微博数+1
    await User.increment({weiboCount: 1}, {where: {id: weibo.userId}});
});
Weibo.afterDestroy(async (weibo) => {
    // 微博数-1
    await User.increment({weiboCount: -1}, {where: {id: weibo.userId}});
});
return Weibo;
};

module.exports.PublishType = {
    Self: 1, // 自己发布
    Share: 2 // 转发
};
```

9.3.7 生成数据表

本项目使用的数据库名称为"weibo",如果该数据库不存在,请使用如下 SQL 命令创建之:

```
CREATE DATABASE weibo;
```

数据表的生成直接使用 sequelize 即可,需要注意的是我们拆分了模型文件,因此需要 import 之后才能建表。

```js
// db.js
const sequelize = require('./shared/sequelize');

sequelize.import('./models/comment');
sequelize.import('./models/user');
sequelize.import('./models/weibo');

sequelize.sync({force: true}).catch((err) => console.error(err)).finally(() =>
    sequelize.close());
```

9.3.8 业务代码

业务代码就是和实际业务息息相关的逻辑,比如用户注册、登录、发表微博等,这些都是业务层的代码。

本项目的业务层代码放在 services 目录中,每个业务模块定义在一个 JS 文件中,每一个业务单元定义在一个独立的函数中。

大家在编写业务代码时,一定要注意检测相关参数,防止不合法请求导致数据出现问题。

1. 评论业务

```js
// services/comment.js
const sequelize = require('../shared/sequelize');
const weiboService = require('./weibo');
const Comment = sequelize.import('../models/comment');
const User = sequelize.import('../models/user');
// 评论微博
exports.publish = async function (weiboId, userId, content) {
    const weibo = await weiboService.show(weiboId);
    if (weibo === null) {
        throw new Error('微博不存在');
    }
    return Comment.create({
        content,
        weiboId,
        userId
    });
};
// 删除评论
exports.destroy = async function (commentId, userId) {
    const comment = await Comment.findByPk(commentId);
    if (comment === null || comment.userId !== userId) {
        throw new Error('你无权删除该评论');
    }
    return comment.destroy();
};
// 查看微博的评论列表
exports.listByWeibo = async function (weiboId, page, size) {
    return Comment.findAndCountAll({
        where: {weiboId},
```

```
        include: [{
            model: User,
            attributes: ['id', 'nickname'],
            as: 'user'
        }],
        offset: (page - 1) * size,
        limit: size,
        order: [['id', 'DESC']]
    });
};
```

2. 用户业务

```
// services/user.js
const sequelize = require('../shared/sequelize');
const User = sequelize.import('../models/user');

// 注册账号
exports.register = async function (username, password) {
    const user = await User.findOne({
        where: {username},
    });
    if (user !== null) {
        throw new Error("账号已存在");
    }
    return User.create({
        username,
        password
    });
};
// 登录
exports.login = async function username(username, password) {
    const user = await User.findOne({
        where: {username}
    });
    if (user === null || !user.checkPassword(password)) {
        throw new Error('账号或密码错误');
    }
    return user;
};
// 查看用户信息
exports.show = function (userId) {
    return User.findByPk(userId, {
        attributes: ['id', 'nickname', 'weibo_count']
    });
};
// 修改个人资料
exports.changeProfile = function (userId, nickname, password) {
    return User.update({nickname, password: password || ''}, {where: {id: userId},
        individualHooks: true});
};
```

```js
// services/weibo.js
const sequelize = require('../shared/sequelize');
const PublishType = require('../models/weibo').PublishType;
const Weibo = sequelize.import('../models/weibo');
const User = sequelize.import('../models/user');
// 发表微博
exports.publish = async function (userId, content) {
    return Weibo.create({
        userId,
        type: PublishType.Self,
        content
    });
};
// 转发微博
exports.share = async function (userId, weiboId, shareContent) {
    const weibo = await Weibo.findByPk(weiboId);
    if (weibo === null) {
        throw new Error('转发的微博不存在');
    }
    if (weibo.userId === userId) {
        throw new Error('不能转发自己的微博');
    }

    return sequelize.transaction(async (transaction) => {
        const newWeibo = await Weibo.create({
            userId,
            content: weibo.content,
            shareContent,
            type: PublishType.Share,
        }, {transaction});
        // 更新源微博的分享数
        weibo.shareCount++;
        await weibo.save({transaction});
        return newWeibo;
    });
};

// 查询一条微博
exports.show = async function (id, withUser = false) {
    const options = {};
    if (withUser) {
        options.include = [{
            model: User,
            as: 'user'
        }];
    }
    return Weibo.findByPk(id, options);
};

// 所有微博列表
```

```js
exports.list = async function (page = 1, size = 10) {
    return Weibo.findAndCountAll({
        limit: size,
        offset: (page - 1) * size,
        order: [['id', 'DESC']],
        include: [
            {
                model: User,
                attributes: ['id', 'nickname'],
                as: 'user'
            },
        ]
    });
};

// 用户发表的微博列表
exports.listByUser = async function (userId, page = 1, size = 10) {
    return Weibo.findAndCountAll({
        where: {userId},
        limit: size,
        offset: (page - 1) * size,
        order: [['id', 'DESC']],
    });
};

// 编辑微博
exports.update = async function (id, userId, content) {
    const weibo = await Weibo.findByPk(id);
    if (weibo === null || weibo.userId !== userId) {
        throw new Error('您无权编辑该微博');
    }
    if (weibo.type !== PublishType.Self) {
        throw new Error('只能编辑自己发布的微博');
    }

    weibo.content = content;
    return weibo.save();
};
// 删除微博
exports.destroy = async function (id, userId) {
    const weibo = await Weibo.findByPk(id);
    if (weibo === null || weibo.userId !== userId) {
        throw new Error('您无权删除该微博');
    }
    return weibo.destroy();
};
```

9.3.9 路由代码

路由层主要完成输入数据的校验以及渲染处理结果的页面。

1. 评论模块

删除评论需要登录才可以，因此我们在路由级别导入 guard 中间件。

ctx.redirect('back')为返回上一页。

```javascript
// routes/comment.js
const Router = require('koa-router');
const commentService = require('../services/comment');
const guard = require('../middlewares/guard');
const router = new Router({prefix: '/comment'});

router.get('/delete/:id', guard, async (ctx) => {
    const userId = ctx.state.userId;
    await commentService.destroy(ctx.params.id, userId);
    await ctx.redirect('back');
});

module.exports = router;
```

2. 广场模块

广场模块是网站的默认首页，以时间顺序显示所有人最新发布的微博以及对应的发布人信息。

```javascript
// routes/home.js
const Router = require('koa-router');
const router = new Router();
const weiboService = require('../services/weibo');

router.get('/', async (ctx) => {
    let {page = 1, size = 10} = ctx.query;
    page = Number(page);
    size = Number(size);
    const {rows, count} = await weiboService.list(page, size);
    await ctx.render('home', {
        list: rows,
        count,
        page: page,
        size: size
    });
});

module.exports = router;
```

3. 用户模块

用户模块负责用户相关业务，在需要登录才能请求的路由中使用了 guard 中间件。

```javascript
// routes/user.js
const Router = require('koa-router');
const userService = require('../services/user');
const weiboService = require('../services/weibo');
const guard = require('../middlewares/guard');
const router = new Router({prefix: '/user'});

router.get('/login', async (ctx) => {
    await ctx.render('user/login');
});

router.post('/login', async (ctx) => {
    const {username, password} = ctx.request.body;
    if (!username || !password) {
        throw new Error('请填写完整!');
    }
    const user = await userService.login(username, password);
    ctx.cookies.set('userId', user.id, {
        signed: true,
        maxAge: 3600 * 24 * 1000
    });
    await ctx.redirect('/');
});

router.get('/register', async (ctx) => {
    await ctx.render('user/register');
});

router.post('/register', async (ctx) => {
    const {username, password, confirmPassword} = ctx.request.body;
    if (!username || !password || !confirmPassword) {
        throw new Error('请填写完整!');
    }
    if (password !== confirmPassword) {
        throw new Error('确认密码不一致');
    }
    await userService.register(username, password);
    await ctx.redirect('/user/login');
});

router.get('/logout', async (ctx) => {
    ctx.cookies.set('userId', null, {maxAge: 0});
    await ctx.redirect('/');
});

router.get('/home', guard, async (ctx) => {
    const {page = 1, size = 10} = ctx.query;
    const {rows, count} = await weiboService.listByUser(ctx.state.userId, page,
```

```
        size);
    await ctx.render('user/home', {
        list: rows,
        count,
        page: Number(page),
        size: Number(size)
    });
});
router.get('/homepage/:id', async (ctx) => {
    const {page = 1, size = 10} = ctx.query;
    const {rows, count} = await weiboService.listByUser(ctx.params.id, page,
     size);
    await ctx.render('user/homepage', {
        list: rows,
        count,
        page: Number(page),
        size: Number(size)
    });
});
router.get('/profile', guard, async (ctx) => {
    const user = await userService.show(ctx.state.userId);
    await ctx.render('user/profile', {
        user
    });
});
router.post('/profile', guard, async (ctx) => {
    const {nickname, password} = ctx.request.body;
    if (!nickname || nickname.length > 20) {
        throw new Error('昵称不合法');
    }
    await userService.changeProfile(ctx.state.userId, nickname, password);
    await ctx.redirect('/user/home');
});
module.exports = router;
```

4. 微博模块

微博模块是本次业务的核心部分,我们一起来看看。

```
// routes/weibo.js
const Router = require('koa-router');
const weiboService = require('../services/weibo');
const commentService = require('../services/comment');
const guard = require('../middlewares/guard');
const router = new Router({prefix: '/weibo'});

router.post('/publish', guard, async (ctx) => {
    const {content} = ctx.request.body;
    if (!content) {
        throw new Error('微博内容不能为空');
```

```js
    }
    if (content.length > 140) {
        throw new Error('微博最长140字');
    }
    await weiboService.publish(ctx.state.userId, content);
    await ctx.redirect('/');
});

router.get('/edit/:id', guard, async (ctx) => {
    const weibo = await weiboService.show(ctx.params.id);
    if (!weibo || weibo.userId !== ctx.state.userId) {
        throw new Error('微博不存在');
    }
    await ctx.render('weibo/edit', {
        weibo
    });
});

router.post('/edit/:id', guard, async (ctx) => {
    const {content} = ctx.request.body;
    if (!content) {
        throw new Error('微博内容不能为空');
    }
    if (content.length > 140) {
        throw new Error('微博最长140字');
    }
    await weiboService.update(ctx.params.id, ctx.state.userId, content);
    await ctx.redirect('back');
});

router.get('/delete/:id', guard, async (ctx) => {
    await weiboService.destroy(ctx.params.id, ctx.state.userId);
    await ctx.redirect('back');
});

router.get('/show/:id', async (ctx) => {
    let {page = 1, size = 10} = ctx.query;
    page = Number(page);
    size = Number(size);
    // 读取微博
    const weiboId = ctx.params.id;
    const weibo = await weiboService.show(weiboId, true);
    if (!weibo) {
        throw new Error('微博不存在');
    }
    // 读取评论
    const {rows: comments, count} = await commentService.listByWeibo(weiboId, page, size);
    await ctx.render('weibo/show', {
        weibo,
        comments,
```

```js
        count,
        page,
        size
    });
});

router.post('/comment/:id', guard, async (ctx) =>{
    const {content} = ctx.request.body;
    if (!content || content.length > 140) {
        throw new Error('评论内容长度不合法');
    }

    const weiboId = ctx.params.id;
    const weibo = await weiboService.show(weiboId);
    if (!weibo) {
        throw new Error('微博不存在');
    }
    await commentService.publish(weiboId, ctx.state.userId, content);
    await ctx.redirect(`/weibo/show/${weiboId}`);
});

router.get('/share/:id', guard, async (ctx) => {
    const weibo = await weiboService.show(ctx.params.id, true);
    if (!weibo) {
        throw new Error('微博不存在');
    }
    const userId = ctx.state.userId;
    if (weibo.userId === userId) {
        throw new Error('不能转发自己的微博');
    }
    await ctx.render('weibo/share', {
        weibo
    });
});

router.post('/share/:id', guard, async (ctx) => {
    const {content} = ctx.request.body;
    if (!content || content.length > 140) {
        throw new Error('转发内容长度不合法');
    }
    const weibo = await weiboService.share(ctx.state.userId, ctx.params.id,
     content);
    await ctx.redirect(`/weibo/show/${weibo.id}`);
});
module.exports = router;
```

9.3.10 视图文件

本项目视图层采用了 AmazeUI 框架,视图代码需要一定的 HTML 知识和 JavaScript 知识,由于篇幅限制，各位读者可以前往 GitHub 仓库查看本项目的视图文件。

9.3.11 Web 应用引导文件

引导文件需要组装中间件和路由，并启动 Web 服务器。

```javascript
// index.js
const Koa = require('koa');
const render = require('koa-ejs');
const bodyParser = require('koa-bodyparser');
const staticMiddleware = require('koa-static');
const errorHandler = require('./middlewares/errorHandler');
const authenticate = require('./middlewares/authenticate');
const homeRoute = require('./routes/home');
const userRoute = require('./routes/user');
const weiboRoute = require('./routes/weibo');
const commentRoute = require('./routes/comment');

const app = new Koa({
    keys: ['KGJ6NLxqOkYCNr1h']
});

// 中间件
app.use(errorHandler);
render(app, {                      // 使用 ejs 中间件
    root: './templates',           // 模板目录
    layout: 'layout',              // 关闭模板布局
    viewExt: 'ejs'
});
app.use(staticMiddleware(__dirname + '/public'));
app.use(bodyParser());
app.use(authenticate);
// 路由
app.use(homeRoute.routes()).use(homeRoute.allowedMethods());
app.use(userRoute.routes()).use(userRoute.allowedMethods());
app.use(weiboRoute.routes()).use(weiboRoute.allowedMethods());
app.use(commentRoute.routes()).use(commentRoute.allowedMethods());

app.listen(8080, () => {
    console.log('listen on 8080')
});
```

根据 Koa 的中间件模型和以往经验，引导文件的模块组装顺序如下：

（1）错误处理器。最先注册的中间件在最外层，因此错误处理器可以捕获所有错误。
（2）第三方中间件。也就是 npm 安装的。
（3）自定义全局中间件。
（4）路由文件。

9.4　效果展示

本项目的主要页面效果展示如下：

首页如图 9-1 所示。

图 9-1

登录页如图 9-2 所示。

图 9-2

注册页如图 9-3 所示。

图 9-3

个人中心如图 9-4 所示。

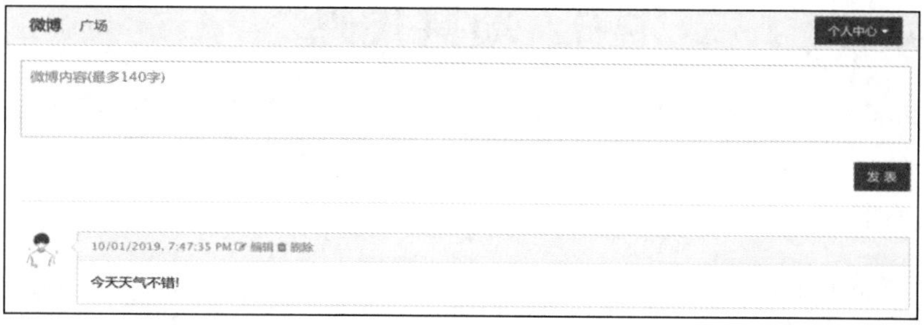

图 9-4

修改资料如图 9-5 所示。

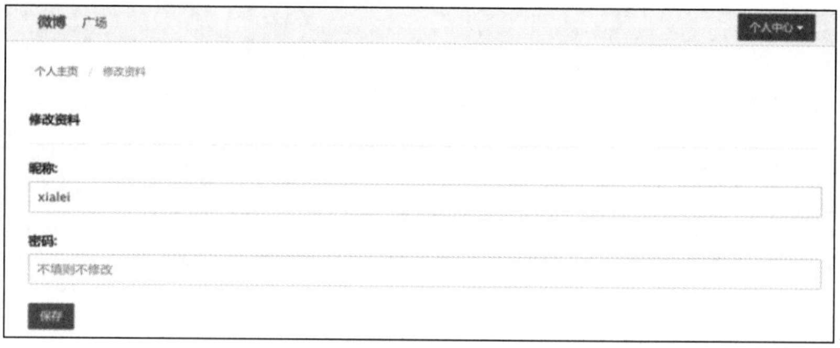

图 9-5

微博详情如图 9-6 所示。

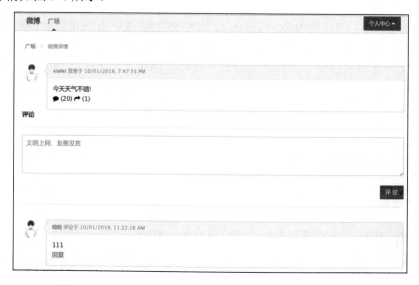

图 9-6

9.5 项目代码

本项目托管于 GitHub，项目网址为：https://github.com/nodejs-inaction/weibo，使用本项目的步骤说明如下。

（1）安装 Node.js 和 MySQL。
（2）克隆代码到本地：

```
git clone
```

（3）安装依赖：

```
npm install
```

（4）同步数据库：

```
node db.js
```

（5）启动项目：

```
node index.js
```

9.6 本章小结

至此，你已经学会了如何使用 koa+sequelize 来开发完整的 Web 应用，包括项目功能分析、结构规划、代码编写，等等。这是一个比较小型但是结构完整的 Web 项目，根据实际需求可以增删组件，各位读者可以基于本项目进行二次开发，增加自己需要的功能，比如关注、点赞、@某人等。

回顾一下本章所学：

- 功能分析。
- 依据功能分析定义数据模型。
- 代码分层架构。
- 项目开发流程。

有任何开发上的问题都可以联系作者公众号或者 GitHub。

第 10 章

高性能内存型 NoSQL 数据库——Redis

本章内容

- Redis 简介和安装
- Redis 常用数据结构
- Node.js 操作 Redis

10.1 Redis 简介

在前面的章节中,我们学习了 NoSQL 文档型数据库 MongoDB 和关系型数据库 MySQL,这两个数据库的主要数据存放在硬盘中,热点数据或者索引等部分数据则存放在内存中。而 Redis 数据库的所有数据都是存放在内存中进行操作的,硬盘更多的是用于存储数据备份,因此 Redis 拥有极高的性能,常用来实现高性能缓存,是构建高性能应用必不可少的组件。

10.1.1 特点

Redis 是完全开源的高性能键-值(Key-Value)数据库。Redis 与经典的 Memcached(一款键-值缓存服务器)相比有以下特点:

- Redis 支持数据持久化,可以将内存中的数据保存到磁盘,重启时可以从磁盘读取到内存,而 Memcached 重启之后数据丢失。
- Redis 支持丰富的数据结构,如 string、list、set、zset、hash 等,而 Memcached 只支持 string。
- Redis 支持主从模式,Memcached 不支持。

- Redis 支持消息订阅模式，Memcached 不支持。

10.1.2 应用场景

Redis 在日常开发中使用场景还是挺多的，大致在以下场景会更适合使用 Redis：

- 热点数据缓存。Redis 访问数据比数据库快几个数量级，因此针对热点数据可以实现缓存，保障访问速度。
- 限时业务。比如手机验证码 60 秒只能发一条的场景。
- 计数器场景。Redis 的原子递增命令可以用于高并发的秒杀活动、接口限流等等。
- 排行榜和投票场景。借助 Redis 的 zset 实现点赞数据实时排行，而数据库实现的效率非常之低。
- 分布式锁。分布式锁一般用在分布式系统中以防止出现高并发导致的数据混乱问题。
- 消息订阅模式。比如推送系统，推送系统的发布者往 Redis 写消息，而消费者监听 Redis 消息完成实际的消息推送。

10.2 Redis 安装

10.2.1 在 Windows 下安装 Redis

Redis 官方提供的是 C 语言源代码，未提供 Windows 版本，幸运的是微软为 Windows 平台提供了编译好的版本，我们可以直接下载下来使用。

下载地址：https://github.com/microsoftarchive/redis/releases。

下载之后解压，在 CMD 窗口中（即命令行窗口中），切换到解压目录下，再执行下列命令以启动 Redis：

```
redis-server.exe redis.windows.conf
```

Redis 服务启动成功之后，再打开另一个 CMD 窗口，执行下列客户端命令，以连接到 Redis 服务器：

```
redis-cli
```

执行成功后就会连接到 Redis 服务器并打开 Redis 终端：

```
redis 127.0.0.1:6379>
```

输入 PING 命令来测试 Redis 服务是否工作正常：

```
redis 127.0.0.1:6379> PING
PONG
```

10.2.2 在 Linux 下安装 Redis

1. Ubuntu

Ubuntu 官方仓库提供了编译好的 Redis 二进制版本，可以直接执行 apt-get 命令来安装：

```
sudo apt-get install redis-server redis-cli
```

安装完成后 Redis 服务器会自动启动。如果需要手动管理，可以在终端输入以下命令：

```
service redis-server {start|stop|restart}
```

服务启动成功后，可以打开客户端工具进行测试：

```
redis-cli
127.0.0.1:6379> PING
PONG
```

2. CentOS

在 CentOS 中使用 Yum 安装 Redis 二进制版本，可以直接执行 yum 命令来安装：

```
yum install redis
```

安装完成后 Redis 服务器会自动启动。如果需要手动管理，可以在终端输入以下命令：

```
systemctl {start|stop|restart} redis
```

服务启动成功后，可以打开客户端工具进行测试：

```
redis-cli
127.0.0.1:6379> PING
PONG
```

10.2.3 在 macOS 下安装 Redis

在 macOS 下，推荐使用 Homebrew 安装 Redis，可以直接执行 brew 命令来安装：

```
brew install redis
```

安装之后需要手动启动 Redis 服务器，可以执行以下命令来启动 Redis 服务器：

```
brew services start redis
```

服务启动成功后，可以打开客户端工具进行测试：

```
redis-cli
127.0.0.1:6379> PING
PONG
```

10.3 Redis 支持的数据结构

Redis 支持 String、Hash、List、Set、ZSet 等数据结构，本节分别讲解一下这些数据结构。

10.3.1 String（字符串）

String 类型用来存储键-值（Key-Value）型数据，一般多用于数据缓存。

```
127.0.0.1:6379> SET name xialei
OK
127.0.0.1:6379> GET name
"xialei"
```

在以上示例中，我们使用 SET 命令设置了一个键为 name，值为 xialei 的元素。接下使用 GET 命令获取该键。

如果把以上示例与 JS 对象类比的话，结果如下：

```
{
  name: "xialei"
}
```

表 10-1 列出了 Redis 常用的字符串命令。

表 10-1　Redis 常用的字符串命令

命令	说明
SET key value	设置指定键的值
GET key	读取指定键的值
MGET key1, key2, ...keyN	一次读取多个键的值
SETEX key seconds value	指定键的值及其过期的时间（单位为秒）
SETNX key value	键不存在时才设置值
MSET key1,value1,...keyN,valueN	一次设置多个键-值对
MSETNX key1,value1,...keyN,valueN	当所有键都不存在时一次设置多对键-值
PSETEX key milliseconds value	指定键的值及其过期的时间（单位为毫秒）
INCRY key	将键存储的数字自增 1
DECRY key	将键存储的数字自减 1
INCRYBY key increment	将键存储的数字自增给定的增量
DECRYBY key decrement	将键存储的数字减去给定的减量

10.3.2 哈希表（Hash）

哈希表是一个键-值的映射表，特别适合存储对象。
哈希表的三个术语如下：

- key：哈希表的名称。
- field：哈希表的字段名。
- value：哈希表的字段值。

比如设置用户 ID 为 1 的姓名和性别到 Redis，命令如下。

```
HMSET user-1 name "xialei" sex 1
```

以上示例构建了一个包含两个字段的 user-1 对象，其中 user-1 为键，name 为字段 1，"xialei" 为值 1，sex 为字段 2，1 为值 2。

如果把以上示例类比于 JavaScript 语句，结果如下：

```
{
  'user-1': {
    name: "xialei",
    sex: 1
  }
}
```

表 10-2 列出了 Redis 常用的哈希命令。

表 10-2　Redis 常用的哈希表命令

命令	说明
HDEL key field1 ...fieldN	删除哈希表的多个字段
HEXISTS key field	检查哈希表的字段 field 是否存在
HGET key field	获取哈希表指定的字段值
HGETALL key	获取哈希表所有字段和值
HINCRBY key field increment	为哈希表指定字段的数字加上增量
HKEYS key	获取哈希表所有字段的列表
HLEN key	获取哈希表字段的数量
HMGET key field1 ...fieldN	读取哈希表多个字段
HMSET key field1 value1 ...fieldN valueN	批量设置哈希表的多个字段和值
HSET key field value	设置哈希表指定字段的值
HSETNX key field value	当哈希表指定字段不存在时才设置值
HVALS key	获取哈希表所有值的列表

（续表）

命令	说明
HSCAN key cursor [MATCH pattern] [COUNT count]	遍历哈希表中键-值对，一般用在哈希表字段多的情况

10.3.3 列表（List）

列表按照插入顺序排序，可以添加到列表的头部，也可以添加到列表的尾部。

```
127.0.0.1:6379> LPUSH books thinkphp
(integer) 1
127.0.0.1:6379> LPUSH books node.js
(integer) 2
127.0.0.1:6379> LPOP books
"node.js"
127.0.0.1:6379> LPOP books
"thinkphp"
```

在以上示例中，我们使用 LPUSH 命令往 books 列表插入了两个元素，然后使用 LPOP 命令弹出了两个元素。

如果把以上示例比作 JavaScript 语句，结构如下：

```
{
  books: ['thinkphp','node.js']
}
```

表 10-3 列出了 Redis 常用的列表命令。

表 10-3　Redis 常用的列表命令

命令	说明
BLPOP key timeout	移除并获取列表的第一个元素。如果没有元素，则阻塞到有元素或发生超时
BRPOP key timeout	移除并获取列表的最后一个元素。如果没有元素，则阻塞到有元素或发生超时
LINDEX key index	通过下标获取列表元素
LLEN key	获取列表长度
LPOP key	移除并获取列表第一个元素，该方法不会阻塞
LPUSH key value ... valueN	将一个或多个值插入列表头部
LRANGE key start stop	获取列表指定范围内的元素
LREM key count value	移除指定数量的列表元素
LSET key index value	通过下标设置列表元素的值
RPOP key	移除并获取列表最后一个元素，该方法不会阻塞
RPUSH key value ... valueN	将一个或多个值插入列表尾部

10.3.4 集合（Set）

Redis 的集合是无序集合，集合中的元素是唯一的。

```
127.0.0.1:6379> SADD books thinkphp thinkphp node.js
(integer) 1
127.0.0.1:6379> SCARD books
(integer) 2
```

在以上示例中，我们使用 SADD 命令向 books 插入了 3 个元素，不过有两个重复元素，所以使用 SCARD 命令获取元素个数时返回 2。

表 10-4 列出了 Redis 常用的集合命令。

表 10-4　Redis 常用的集合命令

命令	说明
SADD key value1 ... valueN	向集合添加一个或多个元素
SCARD key	获取集合元素的个数
SISMEMBER key value	判断指定值 value 是否在集合 key 中
SMEMBERS key	返回集合中所有的元素
SPOP key	移除并返回集合中的一个随机元素
SRANDMEMBER key [count]	返回集合中一个或多个元素
SREM key value1 ... valueN	移除集合中一个或多个元素
SSCAN key cursor [MATCH pattern] [COUNT count]	遍历集合中的元素

10.3.5 有序集合（ZSet）

和普通集合类似，有序集合也不允许重复元素，不过每个元素会关联一个 double 类型的分数，Redis 通过该分数为 ZSet 元素按从小到大排序。ZSet 元素不允许重复，但是分数可以重复。

有序集合特别适合排行榜类的应用，传统的 SQL 数据库实现该类功能时由于涉及统计，性能比较低下，而 Redis 不存在这个问题。

```
127.0.0.1:6379> ZADD rank 100 xialei
(integer) 1
127.0.0.1:6379> ZADD rank 101 zhangsan
(integer) 1
127.0.0.1:6379> ZADD rank 99 lisi
(integer) 1
127.0.0.1:6379> ZRANGE rank 0 10 1
(error) ERR syntax error
```

```
127.0.0.1:6379> ZRANGE rank 0 10 WITHSCORES
1) "lisi"
2) "99"
3) "xialei"
4) "100"
5) "zhangsan"
6) "101"
```

在以上示例中，我们使用 ZADD 向 rank 有序集合插入了 3 个元素并关联相应的分数，使用 ZRANGE 查看元素列表时，可以发现返回的数据已经排好序了。

表 10-5 列出了 Redis 有序集合的常用命令。

表 10-5　Redis 有序集合的常用命令

命令	说明
ZADD key score1 value1 ...scoreN valueN	向有序集合添加一个或多个元素，或者更新已存在元素的分数
ZCARD key	获取有序集合元素的个数
ZCOUNT key min max	计算有序集合在指定分数区间的元素的个数
ZINCRBY key increment value	对有序集合指定元素的分数加上增量
ZRANGE key start stop [WITHSCORES]	通过下标区间返回有序集合元素的列表
ZRANGEBYSCORE key min max [WITHSCORES] [LIMIT]	通过分数区间返回指定数量的有序集合元素之列表
ZRANK key value	获取有序集合中指定元素的下标
ZREM key value ...valueN	删除有序集合中一个或多个元素
ZREMRANGEBYRANK key start stop	移除有序集合中给定的下标区间的所有元素
ZREMRANGEBYSCORE key min max	移除有序集合中给定的分数区间内的所有元素
ZREVRANGE key start stop [WITHSCORES]	返回有序集合中指定下标区间内的元素，分数从高到低
ZREVRANGEBYSCORE key max min [WITHSCORES]	返回有序集合中指定分数区间内的元素，分数从高到低排序
ZREVRANK key member	返回有序集合中指定元素的排名，按分数从大到小排序
ZSCORE key member	返回有序集合中元素的分数
ZSCAN key cursor [MATCH pattern] [COUNT count]	遍历有序集合中的元素（包括元素和元素分数）

10.3.6　发布订阅

发布订阅是一种消息通信模式：发送者发送消息，订阅者接收消息。

Redis 中通过 PUB/SUB 来实现发布订阅模式，Redis 客户端可以订阅任意数量的频道。

下面是一个发布订阅的示例。首先启动一个终端,执行下列命令订阅 chat 频道:

```
127.0.0.1:6379> SUBSCRIBE chat
Reading messages... (press Ctrl-C to quit)
1) "subscribe"
2) "chat"
3) (integer) 1
```

启动一个新终端,执行下列命令向 chat 频道发布消息:

```
127.0.0.1:6379> PUBLISH chat "nihao"
(integer) 1
```

此时终端 1 将收到以下消息:

```
1) "message"
2) "chat"
3) "nihao"
```

表 10-6 列出了 Redis 发布订阅的常用命令。

表 10-6　Redis 发布订阅的常用命令

命令	说明
PSUBSCRIBE pattern ...patternN	订阅一个或多个符合 pattern 的频道
PUBLISH channel message	向指定频道发送消息
PUNSUBSCRIBE [pattern ... patternN]	退订指定模式的频道
SUBSCRIBE channel ... channelN	订阅一个或多个指定名称的频道
UNSUBSCRIBE [channel ... channelN]	退订指定名称的频道

提醒大家一下,Redis 的发布订阅模式实现得比较简单,如果对该模式有性能或者数据安全等方面的要求,请使用专业的消息队列软件。

10.4　Node.js 集成 Redis

Node.js 使用 redis 模块来实现 Redis 服务,直接使用 npm 安装即可。

```
npm install redis
```

10.4.1　快速开始

安装好 redis 模块即可使用。

```
const redis = require("redis");
const client = redis.createClient();
```

```
client.on("error", function (err) {
    console.log("Error " + err);
});

client.set("string key", "string val", redis.print);
client.hset("hash key", "hashtest 1", "some value", redis.print);
client.hset(["hash key", "hashtest 2", "some other value"], redis.print);
client.hkeys("hash key", function (err, replies) {
    console.log(replies.length + " replies:");
    replies.forEach(function (reply, i) {
        console.log("    " + i + ": " + reply);
    });
    client.quit();
});
```

在以上示例中，我们连接 Redis 之后对集合和哈希表进行了一些操作，操作完毕后退出了客户端，这是一个比较完整的操作流程，不过在实际开发中一般会把 Redis 客户端单独作为一个模块，在需要的地方导入。

redis.createClient 方法创建一个 Redis 客户端，该方法接收一个可选的参数配置连接信息。

- host：Redis 主机。默认 localhost。
- port：Redis 端口。默认 637。
- string_numbers：是否将 Redis 字符串转化为 JS 数字。
- return_buffers：是否返回缓冲区，默认返回的是字符串。
- password redis：连接密码。
- db redis：数据库编号。
- prefix key：前缀。

10.4.2 Promise

Redis 支持的回调函数也是 Node.js 风格的，因此我们可以使用 Bluebird 来处理 Redis。

```
const redis = require('redis');
const bluebird = require('bluebird');

bluebird.promisifyAll(redis.RedisClient.prototype);
bluebird.promisifyAll(redis.Multi.prototype);
```

使用 Promise 和 async/await 的示例如下。

```
const redis = require('redis');
const bluebird = require('bluebird');

bluebird.promisifyAll(redis.RedisClient.prototype);
bluebird.promisifyAll(redis.Multi.prototype);
const client = redis.createClient();
```

```
async function example() {
  const data = await client.setAsync("string key", "string val");
  const data1 = await client.hsetAsync("hash key", "field1", "value1");
  const data2 = await client.hkeysAsync("hash key");
  data2.forEach((item, i) => {
    console.log(i+":"+item);
  });
  client.quit();
}
```

其他操作方法名称和参数与 Redis 常用命令类似，有需要的读者可以查看 redis 模块的官方文档。

10.5 本章小结

本章我们学习了 Redis 服务器的基本知识、常用命令和 Node.js 对 Redis 的常用操作。Redis 对于高性能应用是极为重要的一个组件，建议各位读者多花时间学习。

第 11 章

实时双向 Web 技术——WebSocket

本章内容

- 传统的实时 Web 技术
- WebSocket 协议原理
- Node.js 实现 WebSocket 服务器
- 项目实战——公共聊天室

11.1 传统的实时 Web 技术

"实时"Web 技术在很多年前就提出来了，请注意，这里添加了双引号，因为并不是真的实时。HTTP 协议在设计之初就是请求-响应形式的，只有客户端请求了服务器，服务器才可以返回数据，不可以主动推送数据给客户端。对于一些实时性要求很高的应用，比如股票、比赛等应用，HTTP 协议的弊端就暴露出来了。

下面我们看看 WebSocket 出现之前此类问题是如何解决的。

11.1.1 Ajax 轮询（Ajax Polling）

客户端周期性地向服务器发起 Ajax（异步的 HTTP 请求），服务器的响应一般分为以下两种：

```
{
"errcode":1,
"errmsg":"暂无数据"
```

```
}
```

和

```
{
  "errcode": 0,
  "errmsg":"ok",
  data:{ ... }
}
```

每个 Ajax 请求服务器不管有没有数据都会响应，这种模式的缺点很明显，浏览器需要不断地向服务器发出请求，而 HTTP 请求在每次发送时都会带上很长的 HTTP 请求报头，真正有意义的可能只有 URL 部分，这显然会浪费服务器的处理资源。

11.1.2　服务器推送（Comet）

客户端发起 HTTP 请求之后，服务端会将此次请求挂起，当服务器得到最新数据时，会向客户端传输数据，当没有数据更新时，服务器将等待有数据之后才向客户端传送数据。如果服务端长时间没有数据，客户端请求就会超时，此时客户端会重新发起请求。

该模式虽然比 Ajax 轮询要好一点，但是如果发生超时，客户端依旧会产生多个请求，并没有从根本上解决问题。

11.2　WebSocket

WebSocket 同 HTTP 一样是应用层的协议，底层都是 TCP 协议。HTTP 协议是单向通信的，只有客户端发起请求，服务端才能响应。而 WebSocket 协议是双向的，在客户端与服务端建立连接之后，客户端和服务端都可以主动向对方发送数据。WebSocket 需要 HTTP 协议来完成握手操作，握手结束之后双方使用 WebSocket 通信，与 HTTP 协议再无关系。

WebSocket 的协议只有在支持 HTML5 规范的浏览器上才能使用，WebSocket 协议使用的协议头为 ws:// 和 wss://，前者对应 http://，后者对应 https://。

协商过程

WebSocket 是独立的、创建在 TCP 上的协议。WebSocket 通过 HTTP/1.1 协议的 101 状态码进行握手。

为了创建 WebSocket 连接，需要通过浏览器发出请求，之后服务器进行回应，这个过程通常称为"握手"（Handshaking）。握手完毕后复用 HTTP 底层的 TCP 连接，只不过数据格式由 HTTP 格式升级为 WebSocket 格式。

客户端请求报文：

```
GET / HTTP/1.1
```

```
Upgrade: websocket
Connection: Upgrade
Host: example.com
Origin: http://example.com
Sec-WebSocket-Key: sN9cRrP/n9NdMgdcy2VJFQ==
Sec-WebSocket-Version: 13
```

服务端响应报文：

```
HTTP/1.1 101 Switching Protocols
Upgrade: websocket
Connection: Upgrade
Sec-WebSocket-Accept: fFBooB7FAkLlXgRSz0BT3v4hq5s=
Sec-WebSocket-Location: ws://example.com/
```

字段说明：

- Connection：必须设置为 Upgrade，表示客户端希望升级连接。
- Upgrade：必须设置为 websocket，表示希望升级到 WebSocket 协议。
- Sec-WebSocket-Key：是随机的字符串，服务器端会用这些数据来构造出一个 SHA-1 的哈希值，把 Sec-WebSocket-Key 的值加上一个固定的字符串，计算 SHA-1 哈希值，之后进行 Base64 编码，将结果作为"Sec-WebSocket-Accept"响应报头发送给客户端。通过该操作可以防止某些 HTTP 请求也含有这些请求报头导致被误认为 WebSocket 协议。
- Sec-WebSocket-Version：表示 WebSocket 版本。RFC6455 要求使用的版本为 13。
- 其他 HTTP 标准字段，如 Cookie 也可以在 WebSocket 中使用。

浏览器支持：

- IE 10+
- Firefox 11+
- Chrome 16+
- Safari 6+
- Android Webview 4.4+

11.3 实现 WebSocket 握手协议

WebSocket 协议的握手请求是普通的 HTTP 请求，因此我们可以通过 Node.js 来按照规范实现握手请求并承载 WebSocket 协议。

WebSocket 的协议规范文档在网址：https://tools.ietf.org/html/rfc6455。

11.3.1 握手协议过程

（1）客户端生成随机字符串并进行 Base64 编码后作为"Sec-WebSocket-Key"请求报头发送给服务器。

（2）服务端接收"Sec-WebSocket-Key"后与固定密钥"258EAFA5-E914-47DA-95CA-C5AB0DC85B11"连接，该密钥是 WebSocket 协议规定的。

（3）服务端计算连接后字符串的 SHA1 哈希值并进行 Base64 编码。

（4）服务端将该编码作为"Sec-WebSocket-Accept"响应报头发送给客户端。

（5）客户端校验收到的"Sec-WebSocket-Accept"响应报头是否与本地计算出的一致，若一致则升级为 WebSocket 连接。

11.3.2 服务端代码

我们使用 Node.js 来完成握手协议，握手完成之后将进入普通的 TCP 通信，涉及复杂的协议解析，我们在本示例中并没有实现。

```javascript
// 服务端代码
const http = require('http');
const crypto = require('crypto');
// 创建 HTTP 服务器
const server = http.createServer((req, resp) => {
    console.log(`${req.method} ${req.url}`);
    resp.end();
});
// 监听升级事件
server.on('upgrade', (req, socket) => {
    // 读取请求报头的 sec-websocket-key，并和固定密钥连接
    const reqKey = req.headers['sec-websocket-key'] +
    "258EAFA5-E914-47DA-95CA-C5AB0DC85B11";
    // 计算 SHA1 哈希值并进行 Base64 编码
    const respKey = crypto.createHash('SHA1').update(reqKey).digest('base64');
    const respData = [
        'HTTP/1.1 101 Switching Protocols',
        'Upgrade: websocket',
        'Connection: Upgrade',
        'Sec-WebSocket-Accept: ' + respKey,
        'Sec-WebSocket-Location: ws://' + req.headers.host,
        '\r\n'
    ];
    socket.write(respData.join("\r\n"));
    socket.on('data', (data) => {
        console.log(data.toString());
    });
});
```

```
server.listen(8080, () => console.log(8080));
```

在上述代码中，我们监听 upgrade 事件并按照规范计算响应报头，在 upgrade 事件中 req 是标准的 HTTP 请求对象，socket 是对应的 TCP 连接实例，当握手完成之后我们需要监听 socket 上的数据来接收 WebSocket 协议发送的数据。

"258EAFA5-E914-47DA-95CA-C5AB0DC85B11" 是固定的字符串，由 WebSocket 的 RFC 文档规定，用来解决非 WebSocket 请求携带了 Sec-WebSocket-Key 请求报头的场景。

upgrade 事件未回调 Response 对象，因此我们需要手动构造 HTTP 响应并发送给客户端。

11.3.3　客户端代码

```
<!DOCTYPE html>
<html lang="en">
<head>
    <meta charset="UTF-8">
    <title>Title</title>
</head>
<body>
<script>
    const ws = new WebSocket('ws://localhost:8080');
    ws.onopen = function () {
        console.log('open');
        ws.send('helloworld');
    };
    ws.onerror = function (e) {
        console.log(e);
    };
    ws.onmessage = function (message) {
        console.log(message)
    };
</script>
</body>
</html>
```

用浏览器执行该 JS 后，浏览器控制台会输出 "open"，证明连接成功，并发送数据给服务端。服务端控制台会打印乱码，因为目前的 WebSocket 协议报文我们尚未进行解析。

11.4　使用 ws 模块开发聊天室

ws 模块是 Node.js 中最受欢迎的 WebSocket 服务器模块，我们结合 Koa 提供的底层 HTTP 服务器来使用。

11.4.1 安装依赖

```
npm install koa koa-static ws --save
```

11.4.2 服务端代码

```js
// index.js
const Koa = require('koa');
const staticMiddle = require('koa-static');
const WebSocket = require('ws');

const app = new Koa();
app.use(staticMiddle(__dirname + '/public'));

const server = app.listen(8080);
// WebSocket 服务器直接使用 Koa 的 HTTP 服务器
const websocketServer = new WebSocket.Server({
    server
});

// 保存已连接的客户端
const clients = new Set();
// 广播消息
function broadcast(message) {
    for (const client of clients) {
        const {ws, address} = client;
        ws.send(message, (err) => {
            if (err) {
                console.log(`[Broadcast] ${address} error: ${err.message}`);
                clients.delete(client);
                broadcast(`${address} 离开了`);
            }
        });
    }
}
// 监听 WebSocket 事件
websocketServer.on('connection', (ws, request) => {
    const address = request.connection.remoteAddress + ':' +
    request.connection.remotePort;
    const client = {ws, address};
    clients.add(client);
    broadcast(address + ' 连接了');

    ws.on('message', (message) => {
        broadcast(`${address}: ${message}`);
    });
});
```

11.4.3 客户端代码

```html
<!-- public/index.html -->
<!DOCTYPE html>
<html lang="en">
<head>
    <meta charset="UTF-8">
    <title>Chat</title>
</head>
<p>
    连接状态：<span id="status">连接中</span>
</p>
<div id="list" style="border: 1px solid #eee;height: 200px;overflow-y:
    auto"></div>
<textarea id="content" rows="8" style="width: 100%;outline: 0;margin-top: 10px;"
    placeholder="文明发言"></textarea>
<button id="btn" disabled>发送</button>
<body>
<script>
    const ws = new WebSocket("ws://" + location.host);
    const listEle = document.querySelector('#list');
    const contentEle = document.querySelector('#content');
    const btn = document.querySelector('#btn');
    ws.onopen = function () {
        document.querySelector('#status').innerHTML = '已连接';
        btn.disabled = false;
    };
    ws.onmessage = function (message) {
        listEle.innerHTML = listEle.innerHTML + '<br/>' + message.data;
    };
    ws.onerror = function (e) {
        console.log(e);
    };
    ws.onclose = function () {
        document.querySelector('#status').innerHTML = '未连接';
        btn.disabled = true;
    };
    btn.addEventListener('click', function () {
        if (contentEle.value.trim().length === 0) {
            return;
        }
        ws.send(contentEle.value.trim());
        contentEle.value = '';
    }, false);
</script>
</body>
</html>
```

启动服务器之后，打开两个窗口访问 http://localhost:8080 即可实现聊天，如图 11-1 所示。

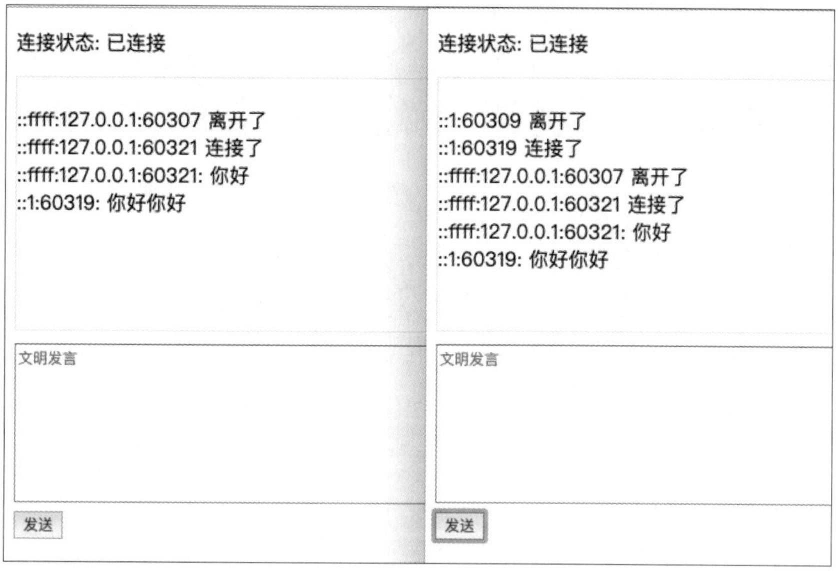

图 11-1

11.5 本章小结

在本章中，我们学习了 WebSocket 的协议规范，使用 Node.js 实现了 WebSocket 握手协议，最后使用 ws 模块实现了一个聊天室。

在实际的应用中，一般会传递 JSON 格式的消息体，消息中包含类型字段，以使接收者可以根据当前所确定的消息类型来做出相应的响应。

Node.js 的后端应用就介绍到这里了，在后端应用部分，我们学习了 Web 框架、数据库操作和缓存操作，这些组件目前已经可以支撑一个常规的 Web 应用了，接下来的章节开始介绍 Node.js 在前端的应用。

第三部分　前端中的Node.js

由于降低了语言门槛，越来越多的前端开发者开始使用Node.js，使得Node.js在前端领域同样大放异彩。在第三部分我们将学习Node.js在前端的应用，包括构建工具、新语言、新框架等等，重点学习的是目前社区最主流的构建工具——Webpack。

第 12 章

迅速发展的前端技术

本章内容

- 模块系统
- 新语言
- 新框架
- 构建工具的发展

随着互联网的加速发展与新标准、新规范的落地，近年来 Web 应用变得更加复杂，Web 技术的应用也更加广泛。从复杂的业务管理后台，到性能要求苛刻的移动端 Web，再到 ReactNative、Flutter 等原生应用解决方案，Web 前端获得了前所未有的机遇，同时也面临更大的挑战。传统的 JavaScript+CSS+HTML 开发 Web 网站的手段已经无法解决 Web 前端面临的问题和挑战。因此，近几年来前端社区出现了许多新语言、新框架和新技术，下面我们一起来看看。

12.1 模块系统

模块化的开发方式可以提高代码复用率，方便进行代码的组织。通常一个 JS 文件就是一个模块，有独立的作用域，只向外部暴露允许的变量、函数和类。

以前开发网页需要通过全局命名空间的方式来组织代码，比如经典的 jQuery 将它的 API 都暴露到了 window.$ 下，其他代码需要$.API 的方式进行调用。这样的组织方式虽然带来了一定的便利性，但更多的是随之而来的问题，包括：

- 命名空间冲突。Zepto 与 jQuery 使用了相同的命名空间 window.$，导致二者无法共存。

- 无法灵活地管理依赖以及依赖的加载顺序。HTML 代码中引用顺序决定了依赖的加载顺序，这在某些场景下是不灵活的。

此外，随着需求的迭代，项目会变得越来越庞大，在不使用模块系统的情况下，项目会变得难以维护。

12.1.1　CommonJS

CommonJS 是一种被广泛使用的 JavaScript 模块化规范，其核心是通过 module.exports 导出需要暴露的接口，通过 require 来加载依赖模块。Node.js 是 CommonJS 规范的主要实践者，Node.js 的流行带来了 CommonJS 的流行，CommonJS 后来被引入到了网页开发中。

采用 CommonJS 模块规范的示例代码如下：

```
// math.js
function sum(a, b) {
  return a+b;
}
// 导出模块
exports.sum = sum;

// index.js
// 加载模块
const sum = require('./math').sum;
```

CommonJS 采用同步的方式加载模块。在服务端，模块文件都保存在本地磁盘，因此加载速度非常快。而在浏览器环境下，由于网络原因，更合理的方式是采用异步模块系统。

12.1.2　AMD

AMD 也是一种 JavaScript 模块化规范，与 CommonJS 最大的不同是它采用了异步的方式去加载依赖模块。RequireJS 是 AMD 模块化规范最具代表性的实现。

采用 AMD 模块规范的示例代码如下：

```
// math.js
// 定义 math 模块
define('math', function() {
  function sum(a, b) {
    return a+b;
  }
  // 导出模块
  return {
    sum: sum
  };
});

// index.js
```

```
require(['math'], function(math) {
  console.log(math.sum(1, 2));
});
```

AMD 解决了浏览器端 CommonJS 存在的问题，支持异步加载和并行加载依赖，但是 JavaScript 运行环境没有原生支持 AMD，需要事先导入支持 AMD 规范的 JS 库。

12.1.3 CMD

CMD 是另一种 JavaScript 模块化规范，与 AMD 规范类似，不同之处在于：

- AMD 推荐依赖前置、提前加载。
- CMD 推荐依赖就近、按需加载。

CMD 规范是 sea.js 在推广过程中产生的。采用 AMD 规范和 CMD 规范的示例代码如下：

```
// AMD 规范，函数执行前加载好依赖
require(['math'], function(math) {
  console.log(math.sum(1, 2));
});

// CMD 规范，运行时加载依赖
define(function(require, exports, module) {
  const math = require('./math');
  console.log(math.sum(1, 2));
});
```

12.1.4 ES6 模块化

ES6 模块化规范是 ECMA（欧洲计算机制造联合会）提出的 JavaScript 模块化规范，它在语言层面规定了模块化。ES6 全称 ECMAScript 6，ECMA 同时也是 ECMAScript 语言规范的制定者。浏览器厂商和 Node.js 都宣布要原生支持该规范，它将逐步取代现存的模块化规范，成为浏览器和服务器通用的模块化解决方案。

JavaScript 的商标目前由 Oracle 公司持有，因此虽然我们使用的时候叫作 JavaScript，但是官方的语言名称叫作 ECMAScript。

采用 ES6 规范的示例代码如下：

```
// math.js
// 导出 sum 函数
export function sum(a, b) {
  return a+b;
}
// index.js
// 加载模块
import {sum} from './math';
console.log(sum(1,2));
```

ES6 模块虽然是目前为止最好的模块化方案，但是目前无法直接运行在 JavaScript 运行环境中，需要通过工具转换为 ES5 代码后才能运行。

12.2 新语言

JavaScript 最初被设计用来做一些简单的功能，比如页面动效、表单处理等等，直接使用 JavaScript 来开发大型应用会有比较大的缺陷，由于语言的问题，这些缺陷可能在运行时才会暴露出来。CSS 最初只能用静态语法描述元素的样式，不可以使用共享、判断等标准编程语法。为了解决以上问题，前端社区诞生了以下新语言：

12.2.1 ES6

ES6（全称 ECMAScript6）是现阶段 JavaScript 的语言标准。它在语言层面为 JavaScript 添加了很多新语法和 API，使得 JavaScript 可以支持大型应用的编写。

新增的特性如下：

- 模块化规范。
- Class 支持。
- let/const 块级作用域支持。
- 箭头函数。
- async/await 语法。
- Symbol。
- 迭代器。

不同的浏览器或服务器程序对规范的实现不一致，为了解决兼容性问题，需要将 ES6 代码编译为 ES5 代码才能运行。Babel 是目前社区最流行的编译工具。

12.2.2 TypeScript

TypeScript 是 JavaScript 的一个超集实现，由 Microsoft 开发，最大的亮点是支持静态类型检查。采用 TypeScript 编写的代码可以被 tsc 编译器编译为 ES5、ES6 等标准的 JavaScript 代码。

静态类型检查可以在开发阶段找出可能存在的问题，提高了应用的健壮性。但是 TypeScript 的语法相对 JavaScript 比较烦琐，并且无法直接在浏览器或服务器运行。

TypeScript 的静态类型检查示例代码如下：

```
function sum(a: number, b: number) {
  return a+b;
}
```

```
// 编译错误,'1'不是数字类型
console.log(sum('1', 2));
```

TypeScript 的官方网站是 https://www.typescriptlang.org/，中文翻译的网站是 https://tslang.cn。

12.2.3 Less

Less 是一种向后兼容的 CSS 扩展语言，可以通过编程的方式编写 CSS。

下面是 Less 语法的示例：

```
@width: 100px;
@height: @width + 10px;
.header {
  width: @width;
  height: @height;
  .header-title {
    color: red;
  }
}
// 编译为 CSS 的结果如下
.header {
  width: 100px;
  height: 110px;
}
.header .header-title {
  color: red;
}
```

Less 提供了变量、函数、混入等特性，不过笔者最喜欢的还是 Less 提供的选择器嵌套语法。

12.2.4 SCSS

SCSS 是另一种 CSS 预处理语言，它支持更加复杂的语法特性。

下面是 SCSS 语法的示例：

```
$width: 100px;
$color: red;
.header {
  width: $width;
  color: $color;
}
// 编译为 CSS 的结果如下
.header {
  width: 100px;
  color: red;
}
```

采用 CSS 预处理语言可以方便地管理代码，提高编码效率。但是，现阶段也需要通过对应的编译器将 CSS 预处理语言转换为标准的 CSS 语言。

12.3 新框架

经典的 jQuery 库通过直接操作 DOM 的方式来编写代码，当 Web 应用变得庞大且复杂时，直接操作 DOM 容易引起性能问题，同时也提高了编程的难度。近年来，前端社区涌现了一大批优秀的前端框架，这里介绍几个比较热门的框架。

在介绍框架之前需要介绍一下一个热门的编程思想——MVVM。

MVVM 可以理解为数据驱动视图的编程思想。在传统的 jQuery 开发中，改变视图需要通过 jQuery 来操作 DOM。而在支持 MVVM 的应用中，直接操作数据后视图会自动更新，开发者只需要专注于数据的处理逻辑而不用关心 DOM 操作。

12.3.1 AngularJS

AngularJS 是 Google 开发的一个 JavaScript 框架。它可通过 script 标签添加到 HTML 页面，它是最早支持 MVVM 编程思想的框架之一。

AngularJS 应用的代码示例如下：

```
<div ng-app>
  <p>你好：{{name}}</p>
  <p>姓名：<input type="text" ng-model="name" /></p>
</div>
```

假设在 input 中输入"demo"时，第一个 p 元素会显示"你好: demo"。DOM 节点内容产生了变化，但是我们只输入了数据，这就是 AngularJS 框架自动完成的功能。

12.3.2 React

React 是 Facebook 开发的框架，创造性地引入了 JSX 语法到 JavaScript 中。可以通过编程方式编写 HTML 代码。

JSX 的示例代码如下：

```
const found = true;
render(found? <p>Found</p>:<p>Not Found</p>);
```

12.3.3 Vue

Vue 是国人开发的一个 MVVM 框架，Vue 将一个组件相关的 HTML 模板、JavaScript 逻辑代码、样式代码都写在一个扩展名为 vue 的文件中，非常方便组件化开发。

```
<template>
<div class="name">hello, {{ name }}</div>
</template>
<script>
  export default {
    data() {
      return {
        name: 'demo'
      }
    }
  }
</script>
<style>
  .name {
    color: red;
  }
</style>
```

12.3.4 Angular

Angular 是 Google 的又一力作，通过 TypeScript 的装饰器语法简化了编程。

```
@Component({
  selector: 'app',
  template: '<h1>{{name}}</h1>'
})
export class AppComponent {
  name = 'demo';
}
```

12.4 构建工具

在前面的内容中我们了解到了前端近年来的新技术、新框架、新语言，但是现阶段，它们都有一个共同点：源代码无法直接运行，必须通过转换后才可以。

构建工具就是用来实现这个功能的，构建工具一般需要提供以下功能：

- 代码转换（必须）：将 ES6 编译成 ES5，将 LESS/SCSS 编译成 CSS。
- 文件优化（必须）：压缩/混淆 JS 代码，压缩 CSS 代码，合并图片等。
- 公共代码提取：提取多个页面公共的 JS 和 CSS 代码，通过浏览器和 HTTP 缓存来减少公共代码的加载。
- 实时刷新：代码变更后能够自动构建、刷新浏览器。

构建工具简化前端开发者重复、机械的劳动，提高了前端开发效率。由于 Node.js 使用

JavaScript 来编写应用，因此大部分构建工具都是基于 Node.js 开发的。我们一起来看看近年来出现的前端构建工具。

12.4.1 Grunt

Grunt 可以自动运行设定的任务，同时由于 Grunt 出现得比较早，因此生态系统非常庞大，拥有数量庞大的插件可供选择，因此可以利用 Grunt 自动完成任何事，并且花费最少的代价。

Grunt 使用 Gruntfile.js 来定义任务和依赖。

```
// Gruntfile.js
module.exports = function(grunt) {
  // 项目配置
  grunt.initConfig({
    pkg: grunt.file.readJSON('package.json'),
    uglify: {
      options: {
        banner: '/*! <%= pkg.name %> <%= grunt.template.today("yyyy-mm-dd") %> */\n'
      },
      build: {
        src: 'src/<%= pkg.name %>.js',
        dest: 'build/<%= pkg.name %>.min.js'
      }
    }
  });

  // 加载包含 "uglify" 任务的插件。
  grunt.loadNpmTasks('grunt-contrib-uglify');

  // 默认被执行的任务列表。
  grunt.registerTask('default', ['uglify']);
};
```

在项目根目录执行 grunt default 即可自动将 src 目录下的 js 文件压缩混淆之后复制到 dest 目录中。

Grunt 的优点是有大量的插件，缺点是需要编写很多配置才可以使用。

12.4.2 Gulp

Gulp 是一个基于流的构建工具。Gulp 通过管道的形式来进行文件转换。Gulp 的 API 非常简单，使用下面的 5 个函数几乎可以支持绝大部分的构建场景。

- gulp.task：注册任务。
- gulp.run：运行任务。
- gulp.watch：监听文件变化并执行操作。
- gulp.src：设置文件来源。

- gulp.dest: 设置文件目的地。

```
const gulp = require('gulp');
const sass = require('gulp-sass');
const jsjint = require('gulp-jshint');
const concat = require('gulp-concat');
const uglify = require('gulp-uglify');

// 编译 scss
gulp.task('scss', function() {
  // scss 源文件
  gulp.src('./src/scss/*.scss')
  // 源文件提供给 gulp-sass 转换为 css
  .pipe(sass())
  // 转换结果输出到 build/css 目录
  .pipe(gulp.dest('./build/css'));
});

// 编译 js
gulp.task('js', function() {
  // js 源文件
  gulp.src('./src/*.js')
  // 合并为一个 js
  .pipe(concat('bundle.js'))
  // 压缩、混淆
  .pipe(uglify())
  // 输出到 build 目录
  .pipe(gulp.dest('./build'));
});

// 监听任务
gulp.task('watch', function() {
  // 监听 ./src/scss 目录下 scss 文件变化，并执行 sass 任务
  gulp.watch('./src/scss/*.scss',['sass']);
  // 监听 ./src 目录下 js 文件变化，并执行 js 任务
  gulp.watch('./src/*.js',['js']);
});
```

Gulp 的优点是简化了 API，同时还有大量的插件可以使用，缺点和 Grunt 类似，也是需要大量的配置才可以使用。

12.4.3　Webpack

Webpack 是一个模块化构建工具，在 Webpack 中一切文件都被视为模块，如果发现不支持的模块类型（原生支持 JS 模块，如果是图片、CSS 等，则不支持），则使用对应的 Loader 进行转换后使用，通过 Plugin 来对构建流程进行操作，最后会输出多个模块文件。

将一切文件都视为模块，可以统一编程方式，也方便 Webpack 进行依赖解析。

Webpack 也是基于配置文件来工作的。

```
module.exports = {
  entry: './src/main',
  output: {
    filename: 'build/main.js'
  }
};
```

Webpack 的优点很多：

- 模块化亲和。
- 通过 Loader 可以将任何不支持的模块转换为受支持的模块。
- 通过 Plugin 可以扩展构建流程。
- 配置简单，开发体验良好，支持开箱即用。

当然，Webpack 有个缺点就是只支持模块化项目，不过现阶段的前端项目来说几乎都是模块化的。

12.5　本章小结

本章我们了解了迅速发展的前端技术，包括模块系统的发展、语言的发展、工具的发展。下一章我们将学习目前主流的构建工具——Webpack。

第 13 章

Webpack 起步

本章内容

- Webpack 的基本配置
- Loader 的使用
- Plugin 的使用

使用 Webpack 执行任务时，需要安装 Webpack 这个软件包，请确保系统已经安装了 Node.js 环境。

Webpack 的安装有全局安装和项目安装两种方式，推荐使用项目安装的方式。

本书中使用的 Webpack 版本为 4.x。

13.1 安 装

（1）新建项目目录，如 app。
（2）进行项目根目录执行 npm init 初始化项目。
（3）执行 npm install webpack --save-dev，将 Webpack 安装到开发依赖。
（4）执行 npm install webpack-cli --save-dev，将 webpack-cli 安装到开发依赖。
（5）编辑 package.json：

```
{
  "scripts": {
    "start":"webpack --config webpack.config.js"
  }
}
```

13.2 示例项目

下面通过 Webpack 来构建一个普通的网页项目。

文件目录如下:

```
|----webpack.config.js
|----build         # 构建结果
|----index.html    入口文件
|----src           # 源文件
    |----main.js   主 JS 文件
```

开始构建前,需要将对应的 JavaScript 文件和 HTML 文件代码编写好。

页面入口文件 src/index.html 代码如下:

```html
<html>
  <head>
    <meta charset="UTF-8">
    <title>Webpack 示例</title>
  </head>
  <button id="btn">点击</button>
  <script src="./build/bundle.js"></script>
</html>
```

执行入口的 main.js 的代码如下:

```js
document.querySelector('#btn').addEventListener('click', function(){
  alert('Hello World');
}, false);
```

业务代码编写完毕,接下来我们编写 webpack.config.js,代码如下:

```js
const path = require('path');
module.exports = {
  entry: './src/main', // 入口文件
  output: {
    // 构建模式设置为开发模式
    mode: 'development',
    // 构建后的文件名
    filename: 'bundle.js',
    // 构建后文件的存放目录
    path: path.resolve(__dirname, './build')
  }
};
```

在项目根目录下执行 npm run start 命令进行构建,执行结束后项目根目录下会自动新建 build 目录,该目录下有 bundle.js,这就是我们构建后生成的文件,它可以直接运行于浏览器环境中。使用浏览器打开 index.html,点击页面上的按钮,将会弹出 Hello World 的信息框。

至此，我们已经学会了 Webpack 的基本功能和项目流程，接下来我们将继续探索 Webpack 的更多知识。

13.3　Loader

在前面的章节中，我们使用 Webpack 构建了一个标准的单页面项目，本节我们将向该页面导入 CSS 和图片。

我们知道，Webpack 将一切文件视为模块，CSS 和图片文件也不例外，因此我们可以在 JS 文件中使用以下代码引用 CSS 文件：

```
require('./main.css');
```

不过上述代码直接运行 Webpack 构建会提示错误，因为 Webpack 原生只支持 JavaScript 文件。如果需要使用非 JavaScript 模块，需要使用 Webpack 的 Loader。

Loader 的主要职责是完成非 JavaScript 模块到 JavaScript 模块的转换。

13.3.1　CSS 处理

Webpack 中使用 CSS 文件需要导入 style-loader 和 css-loader，它们能完成 CSS 文件到 JavaScript 模块的转换。

执行 npm install style-loader css-loader --save-dev 命令安装相关 loader，接下来需要编辑 webpack.config.js，配置解析 CSS 的规则。

```
const path = require('path');

module.exports = {
    mode: 'development',
    entry: './src/main.js',
    output: {
        filename: 'bundle.js',
        path: path.resolve(__dirname, './build')
    },
    module: {
        rules: [
            {
                // 以 css 结尾的文件应用下面的 loader
                test:/\.css$/,
                // 使用 css-loader 和 style-loader
                use: ['style-loader','css-loader']
            }
        ]
    }
};
```

在 src 目录中新建 main.css，内容如下：

```css
#btn {
  color: red;
}
```

编辑 src 目录中的 main.js，导入 main.css，内容如下：

```js
require('./main.css');

document.querySelector('#btn').addEventListener('click', function () {
    alert('Hello World');
}, false);
```

接下来执行构建命令 npm run start，构建完成后打开 index.html 文件，可以发现按钮的文字变成了红色，证明 CSS 应用成功。

这里有的读者可能会有一个疑问，为什么 build 目录没有产生 CSS 文件呢？

这个和我们使用的 Loader 有关系，css-loader 完成 CSS 文件到 JavaScript 模块的转换，而 style-loader 则将 JavaScript 模块中的 CSS 样式内容通过 DOM 操作写入到 HTML 页面的 style 节点中，如果需要单独打包 CSS 文件，则需要配合 Plugin，这个将在后面的章节中学习。

13.3.2 图片处理

图片的处理也需要专门的 Loader，这里我们使用 file-loader 来加载。

执行 npm install file-loader --save-dev 命令来安装 file-loader，安转完毕后编辑 webpack.config.js，配置图片加载规则。

```js
const path = require('path');

module.exports = {
    mode: 'development',
    entry: './src/main.js',
    output: {
        filename: 'bundle.js',
        path: path.resolve(__dirname, './build')
    },
    module: {
        rules: [
            {
                test: /\.css$/,
                use: ['style-loader', 'css-loader']
            },
            {
                // 以 gif/png/jpg 结尾的文件应用 file-loader
                test: /\.(gif|png|jpg)$/,
                use: [
                    {
                        loader: 'file-loader',
```

```
                options: {
                    name: 'images/[name].[ext]',
                    // 构建文件名和目录设置，[name]和[ext]是占位符

                    esModule: false,            // 是否启用 ES6 模块系统，本项目未启用
                    publicPath: 'build',        // 输出的根目录
                }
            }
        ]
    }
}
};
```

Webpack 中 publicPath 默认是/，file-loader 的 name 配置的是 images/[name].[ext]，如果在源代码中我们加载了 logo.gif 图片，那么最终写入到代码中的路径为 images/logo.gif，但是构建结果的根目录是 build，因此代码中需要写入 build/images/logo.gif 才对，这就是 publicPath 的作用。

编辑 index.html，添加一个空的 img 标签，需要指定 ID，因为我们将在 main.js 文件中给该元素指定 src。

```html
<html>
  <head>
    <meta charset="UTF-8">
    <title>Webpack 示例</title>
  </head>
  <button id="btn">点击</button>
  <img id="logo" alt="">
  <script src="./build/bundle.js"></script>
</html>
```

准备一张图片存放在 src 下，图片命名为 logo.gif（笔者准备的是 gif 图片，读者可以依照自己的情况进行命名）。

编辑 main.js，内容如下：

```js
require('./main.css');
// 加载 logo.gif 文件
const logo = require('./logo.gif');

document.querySelector('#btn').addEventListener('click', function () {
    alert('Hello World');
}, false);
// 获取图片元素节点，手动设置图片链接。
document.querySelector('#logo').src = logo;
```

执行 npm run start 构建命令，构建完成后 build 目录的内容如下：

```
|----images
    |----logo.gif
|----bundle.js
```

此时打开 index.html，可以看到图片正常显示了，如图 13-1 所示。

图 13-1

Loader 主要职责就是进行文件转换，通过配置 module.rules 数组，告诉 Webpack 遇到哪些文件使用哪些 Loader 进行加载和转换，使用 Loader 的注意事项如下：

- Loader 的顺序是从后到前的。
- Loader 需要的选项可以通过标准的 JavaScript 对象的形式传入。
- 不同的 Loader 支持的选项是不同的，需要查阅 Loader 对应的文档。

13.4 Plugin

Plugin 的主要职责是扩展 Webpack 功能，通过在构建流程中注入钩子函数来实现。

13.4.1 提取 CSS

在前面的章节中，我们通过 style-loader 将 CSS 文件直接加载到了 HTML 页面上，本节我们使用 Loader 将其单独打包为一个 CSS 文件。

提取 CSS 文件需要使用到 mini-css-extract-plugin 这个 Plugin，需要先安装一下。

```
npm install mini-css-extract-plugin --save-dev
```

安装完毕后编辑 webpack.config.js，使用该 Plugin。

```
const path = require('path');
// 导入插件
const MiniCssPlugin = require('mini-css-extract-plugin');

module.exports = {
    mode: 'development',
    entry: './src/main.js',
    output: {
        filename: 'bundle.js',
        path: path.resolve(__dirname, './build')
    },
    module: {
        rules: [
            {
```

```
                test: /\.css$/,
                // 将 style-loader 替换为 MiniCssPlugin 提供的 loader
                use: [MiniCssPlugin.loader, 'css-loader']
            },
            {
                test: /\.(gif|png|jpg)$/,
                use: [
                    {
                        loader: 'file-loader',
                        options: {
                            name: 'images/[name].[ext]',
                            esModule: false,
                            publicPath: 'build'
                        }
                    }
                ]
            }
        ]
    },
    plugins:[
        // 实例化 MiniCssPlugin
        new MiniCssPlugin({
            // 输出的 CSS 文件名
            filename:'css/[name].css',
        })
    ]
};
```

重新执行构建，我们会发现 build 目录下多出了一个 CSS 目录，该目录下有 main.css 文件，文件内容和 src/main.css 的内容一致，此时我们单独提取 CSS 的工作就完成了。

现在打开 index.html，样式文件是无法生效的，因为我们删除了 style-loader，单独打包了 CSS，而 index.html 没有导入 main.css，我们可以编辑一下 index.html，手动导入 main.css：

```html
<html>
  <head>
    <meta charset="UTF-8">
    <title>Webpack 示例</title>
    <link rel="stylesheet" href="build/css/main.css">
  </head>
  <button id="btn">点击</button>
  <img id="logo" alt="">
  <script src="./build/bundle.js"></script>
</html>
```

此时打开 index.html，可以看到样式正常显示了。

通过 webpack.config.js 的 plugins 配置可以看出，该数组的子元素是独立的插件实例，可以通过构造参数来传入 Plugin 需要的配置属性。例如，MiniCssPlugin 插件的功能是提取 JavaScript 代码中的 CSS 代码到一个单独的文件中。示例中我们传入了 css/[name].css 来告诉插件输出 CSS 的目标路径的规则。

13.4.2 自动更新 HTML 中的资源引用

在前面的内容中，可以发现需要在 index.html 中手动引用构建完成后的 CSS 和 JavaScript 文件，项目简单时手动导入还可以接受，当项目变大或者我们启用 Webpack 的文件哈希功能（自动给文件名添加哈希值解决了浏览器同名文件缓存的问题）时，再通过手动的方式进行引用，不仅效率低，而且容易出错。

下面我们通过 html-webpack-plugin 插件来完成这个功能，使用之前先进行安装：

```
npm run html-webpack-plugin --save-dev
```

安装完成后编辑 webpack.config.js，配置一下 Plugin：

```
const path = require('path');
const MiniCssPlugin = require('mini-css-extract-plugin');
// 导入 HtmlWebpackPlugin
const HtmlWebpackPlugin = require('html-webpack-plugin');

module.exports = {
    mode: 'development',
    entry: './src/main.js',
    output: {
        filename: 'bundle.js',
        path: path.resolve(__dirname, './build')
    },
    module: {
        rules: [
            {
                test: /\.css$/,
                use: [MiniCssPlugin.loader, 'css-loader']
            },
            {
                test: /\.(gif|png|jpg)$/,
                use: [
                    {
                        loader: 'file-loader',
                        options: {
                            name: 'images/[name].[ext]',
                            esModule: false
                        }
                    }
                ]
            }
        ]
    },
    plugins: [
        new MiniCssPlugin({
            filename: 'css/[name].css',
        }),
```

```js
      // 实例化 HtmlWebpackPlugin
      new HtmlWebpackPlugin({
        chunks: ['main'],
        // 该 html 文件引用的 chunks 数组，chunk 的概念将在稍后的内容中介绍
        filename: 'index.html', // 构建后的文件名
        template: 'index.html'  // 源文件名
      })
    ]
};
```

需要注意的是，Webpack 中所有构建输出的文件都会输出到 output.path 中去，因此不需要手动去写 build 目录名称。

接下来编辑 index.html，我们删除所有引用 CSS 和 JavaScript 的代码：

```html
<html>
  <head>
    <meta charset="UTF-8">
    <title>Webpack 示例</title>
  </head>
  <button id="btn">点击</button>
  <img id="logo" alt="">
</html>
```

编辑完毕后执行 npm run start 构建命令，build 目录下将会出现 index.html 文件，内容如下：

```html
<html>
  <head>
    <meta charset="UTF-8">
    <title>Webpack 示例</title>
  <link href="css/main.css" rel="stylesheet"></head>
  <button id="btn">点击</button>
  <img id="logo" alt="">
</html><script type="text/javascript" src="bundle.js"></script>
```

可以看到自动添加上了 CSS 和 JavaScript 文件的引用，路径问题也自动解决了。以后我们在交付时只用交付 build 目录的文件即可。

13.5 开发服务器

在前面的内容中，我们都是编写完代码后再统一进行构建，只要修改了代码就需要重新构建一次和执行一次代码，开发效率是比较低的。

实际上，Webpack 已经提供了一个 webpack-dev-server，启动该服务器时会自动启动一个 HTTP 服务器，同时也会启动一个 Webpack，通过 WebSocket 协议告知浏览器更新文件，以实现实时预览。

执行 npm install webpack-dev-server --save-dev 命令安装 webpack-dev-server，安装完毕后

需要编辑 webpack.config.js，开启 devServer 配置：

```js
const path = require('path');
const MiniCssPlugin = require('mini-css-extract-plugin');
const HtmlWebpackPlugin = require('html-webpack-plugin');

module.exports = {
    mode: 'development',
    entry: './src/main.js',
    output: {
        filename: 'bundle.js',
        path: path.resolve(__dirname, './build')
    },
    module: {
        rules: [
            {
                test: /\.css$/,
                use: [MiniCssPlugin.loader, 'css-loader']
            },
            {
                test: /\.(gif|png|jpg)$/,
                use: [
                    {
                        loader: 'file-loader',
                        options: {
                            name: 'images/[name].[ext]',
                            esModule: false
                        }
                    }
                ]
            }
        ]
    },
    // 开发服务器配置
    devServer: {
        // URL 的根目录
        contentBase: path.resolve(__dirname, 'build'),
        // DevServer 的 HTTP 服务端口
        port: 8080
    },
    plugins: [
        new MiniCssPlugin({
            filename: 'css/[name].css',
        }),
        new HtmlWebpackPlugin({
            chunks: ['main'],
            filename: 'index.html',
            template: 'index.html'
        })
    ]
```

```
};
```

编辑 package.json，添加 dev 启动命令：

```
{
  "scripts": {
  "start": "webpack --config webpack.config.js",
  "dev": "webpack-dev-server"
  }
}
```

webpack-dev-server 会自动读取默认的配置文件名 webpack.config.js，因此 package.json 中写不写没有区别。

运行 npm run dev 命令启动 DevServer，可以看到终端有以下输出：

```
「wds」: Project is running at http://localhost:8080/
「wds」: webpack output is served from /
```

打开浏览器访问 http://localhost:8080 可以看到跟之前一样的输出，不同的是我们目前是通过 HTTP 协议进行访问的，而之前的内容是通过本地文件访问的。

此时可以试着编辑 src 目录中的任何代码，编辑完成后浏览器端将会自动刷新，这个技术被称为 LiveReload 技术。通过 LiveReload 技术，可以有效地提高开发效率，现阶段前端开发一般会配置两个显示器，一个用来编码，一个用来查看效果，有了 LiveReload 之后，可以实时查看编码效果，不需要再手动刷新浏览器。

13.6 核心概念

通过本章前面内容的学习，相信大家对于 Webpack 已经有了一个初步的认识，当然，也会存在一些疑问，前面的内容中有一些新术语和新概念尚未进行完整介绍，下面我们一起来看一下 Webpack 中核心的几个概念：

- Entry：项目入口，Webpack 将从入口文件开始解析依赖关系并进行构建。
- Output：输出配置，Webpack 构建完毕的内容将会输出到配置的目录中。
- Chunk：代码块或代码包，一般由多个模块组成，多用于提取公共模块。
- Loader：模块加载器，用来将 Webpack 不支持的模块转换为 Webpack 支持的模块。
- Plugin：插件，通过在 Webpack 构建流程的钩子函数中插入扩展逻辑，以控制构建流程。

13.7 本章小结

本章我们学习了 Webpack 的基础知识，如何新建一个项目，如何配置 Loader 和 Plugin，以及如何启动开发服务器提高开发效率。下一章我们将详细介绍 Webpack 的配置。

第 14 章

Webpack 配置

本章内容

- Webpack 的配置详解

Webpack 支持两种方式传入配置：

- 通过 JavaScript 文件配置，默认的配置文件名为 webpack.config.js。
- 通过命令行参数传入。

这两种方式可以共存，比如有些参数通过 JavaScript 文件定义，另外一些参数通过命令行传入，实现灵活配置的目的。

Webpack 主要的配置项如下：

- mode：配置应用打包环境。
- entry：配置应用入口。
- output：配置构建输出。
- module：配置模块处理规则。
- resolve：配置寻找模块的规则。
- devtool：配置开发工具。
- context：配置中解析入口起点（Entry）和 loader。
- externals：配置外部运行环境提供的模块。
- devServer：配置开发服务器。
- plugins：配置插件。

14.1 Mode

该选项告知 Webpack 使用相应环境的内置优化。

mode 可以配置为 production（默认）、development 或 none。

可以通过配置文件进行配置：

```
module.exports = {
  mode: 'production'
};
```

也可以通过启动 Webpack 时从命令行传入：

```
webpack --mode=production
```

mode 配置选项如表 14-1 所示。

表 14-1 mode 配置选项

选项	说明
production	会将 DefinePlugin 中 process.env.NODE_ENV 的值设置为 production。启用 FlagDependencyUsagePlugin、FlagIncludedChunksPlugin、ModuleConcatenationPlugin、NoEmitOnErrorsPlugin、OccurrenceOrderPlugin、SideEffectsFlagPlugin 和 TerserPlugin
development	会将 DefinePlugin 中 process.env.NODE_ENV 的值设置为 development。启用 NamedChunksPlugin 和 NamedModulesPlugin
none	不使用任何默认的优化选项

注意，设置 NODE_ENV 环境变量不会自动设置 mode 选项。

14.2 Entry 和 Context

Webpack 通过 Entry 入口来开始应用的构建，Context 是包含入口文件的目录，必须使用绝对路径。

假设我们有以下的应用结构：

```
|----webpack.config.js
   |----src
      |---- main.js
```

14.2.1 不配置 Context 的情况

不配置 Context 的情况下，在哪个目录执行 Webpack 命令，Webpack 就会从该目录进行入口文件的搜索。

配置文件如下：

```
module.exports = {
  // 根目录是执行 Webpack 命令的目录，因此 main.js 相对根目录的路径为 ./src/main.js
  entry: './src/main.js',
};
```

14.2.2 配置 Context 的情况

下面是配置 Context 的情况：

```
module.exports = {
  // main.js 相对应用入口目录 src 的路径为 ./main.js
  entry: './main.js',
  // 指定 src 目录为应用入口目录
  context: path.resolve(__dirname, 'src'),
};
```

以上两种配置都是可以正常工作的。

Entry 支持字符串、对象或数组格式的配置，如表 14-2 所示。

表 14-2　Entry 支持字符串、对象或数组格式的配置

类型	说明	示例
字符串	单个入口文件，打包后的 chunk 名称为 main	'./src/main'
数组	单个入口	['./main','./second']
对象	多个入口文件，打包后的 chunk 名称为对应的键名	{main:'./main', second:'./second'}

14.3　Output

构建的输出配置，只支持对象形式的配置。下面我们一起来看看。

filename

配置输出文件的名称，这些输出文件将写入到 output.path 选项指定的目录下。

对于单个入口的应用，filename 可以指定为固定的值，比如 bundle.js，但是对于多个入口的项目，就需要借助 Webpack 提供的占位符了。比如下面的 Entry 配置了三个入口，输出结果得到三个文件：

```
module.exports = {
  entry: {
    main: './src/main',
    second: './src/second',
```

```
    third: './src/third',
  },
  output: {
    filename: '[name].js',
    path: path.resolve(__dirname, 'build')
  }
};
```

配置中的[name]在最终输出时会被 Webpack 替换为真实的名称，Webpack 常用占位符还有表 14-3 所示的几个。

表 14-3　Webpack 常用占位符

占位符	说明
id	chunk 的 ID，从 0 开始
name	chunk 的名称
hash	项目级的哈希值，不管修改项目的什么文件都会变化
chunkhash	chunk 的哈希值，只有修改该 chunk 内的代码才会变化
contenthash	基于内容的哈希值，只有修改该文件本身才会变化

hash/chunkhash/contenthash 的长度可以自定义，比如[hash:16]代表取 16 位的哈希值。

14.3.1　chunkFilename

该选项指明了非入口的 Chunk 在输出时的文件名称。这些 Chunk 一般是在 Webpack 运行构建过程中产生的，所以 chunkFilename 只对这些运行时生成的 Chunk 生效。chunkFilename 支持和 filename 同样的占位符。

14.3.2　path

该选项配置 Webpack 输出文件的目录，必须使用绝对路径。

```
const path = require('path');

module.exports = {
  output: {
    path: path.resolve(__dirname, 'build')
  }
};
```

上例中 Webpack 会将输出文件写入 build 目录中。

14.3.3 publicPath

该选项把输出目录配置为浏览器环境下的 URL 地址，默认是空字符串，使用相对路径。假设我们构建输出的结构如下：

```
|----build
   |----js
      |----main.js
   |----images
      |----logo.png
   |----css
      |----main.css
   |----index.html
```

生产环境下 JS、CSS、图片资源都会托管到 CDN，但是默认情况下代码中输出的 JS、图片、CSS 地址都是相对于 HTML 文件的地址，publicPath 的作用就在此，比如此次构建后我们线上的 CDN 网址为 https://cdn.ddhigh.com/assets/，那么 build/js/main.js 的 URL 为 https://cdn.ddhigh.com/assets/js/main.js。通过下面的配置可以将 publicPath 配置为指定的地址。

```
module.exports = {
  output:{
    publicPath: 'https://cdn.ddhigh.com/assets/'
  }
};
```

14.3.4 libraryTarget 和 library

当使用 Webpack 去构建一个可以被其他模块导入的库时，需要用到这两个配置，比如你开发了一个弹出框插件，想公开给其他人使用。

- libraryTarget 配置导出库的方式。比如是否使用模块系统，使用什么模块系统，等等。
- library 配置导出库的名称。

libraryTarget 的取值是 Webpack 预定义的，支持以下取值。

1. var（默认）

将库通过 var 赋值给 library 选项定义的变量。

比如定义 library 为 'MyLibrary'，则输出和使用的代码如下：

```
// Webpack 输出的代码
var MyLibrary = code;
// 用户使用库的代码
MyLibrary.func();
```

code 是被导出库的内容，是一个有返回值的立即执行函数。

2. this

将库赋值给 this 对象的属性，属性名由 library 指定。

```
// Webpack 输出的代码
this['MyLibrary'] = code;
// 用户使用库的代码
this.MyLibrary.func();
```

3. window

将库赋值给 window 对象的属性，属性名由 library 指定。该选项和 this 类似。

```
// Webpack 输出的代码
window['MyLibrary'] = code;
// 用户使用库的代码
window.MyLibrary.func();
```

4. global

将库赋值给 window 对象的属性，属性名由 library 指定。该选项和 this 类似。

在浏览器环境中 global 和 window 是同一个对象，在 Node.js 环境中 global 是 Node.js 的全局对象。

```
// Webpack 输出的代码
global['MyLibrary'] = code;
// 用户使用库的代码
global.MyLibrary.func();
```

5. commonjs

commonjs 标准中使用 exports.xxx 来导出变量、函数等等。因此本选项将库赋值给 exports 对象的属性，属性名由 library 指定。

```
// Webpack 输出的代码
exports['MyLibrary'] = code;
// 用户使用库的代码
require('MyLibrary').func();
```

6. commonjs2

commonsj2 标准中使用 module.exports=xxx 来导出对象。因此本选项将库赋值给 module.exports 对象，library 配置将被忽略。由于 library 被忽略，因此使用时直接调用库的 npm 名称即可。

```
// Webpack 输出的代码
module.exports = code;
// 用户使用库的代码
require('库的npm包名').func();
```

7. amd

将库通过 amd 模块系统提供的 define 函数进行定义，定义的模块名称由 library 指定。

```
// Webpack 输出的代码
define('MyLibrary',[], function() {
```

```
  return code;
});
// 用户使用库的代码
require(['MyLibrary'], function(MyLibrary) {
  MyLibrary.func();
});
```

8. umd

将库暴露为所有的模块定义下都可运行的方式，支持 commonjs、AMD 模块系统，或者将模块导出到 global 下的变量。

```
// webpack 自动判断模块系统的绝对导出方式
(function webpackUniversalModuleDefinition(root, factory) {
  if(typeof exports === 'object' && typeof module === 'object')
    module.exports = factory();
  else if(typeof define === 'function' && define.amd)
    define([], factory);
  else if(typeof exports === 'object')
    exports['MyLibrary'] = factory();
  else
    // 无可用模块系统，导出到 global
    root['MyLibrary'] = factory();
})(typeof self !== 'undefined' ? self : this, function() {
  return code;
});
```

14.4　Module

本节详细讲解 module 配置模块处理规则。

14.4.1　noParse

noParse 配置用来告知 Webpack 忽略指定的、且未采用模块系统的模块，不对其进行递归解析和处理。这样做的好处是可以提高构建性能，因为一些大型的库（如 jQuery）没有采用模块系统，让 Webpack 解析浪费时间和资源。

noParse 支持的一些配置如下：

```
// 以正则表达式方式指定
noParse: /jquery/
// 以数组方式指定
noParse: [/jquery/]
// 以函数方式指定
noParse: (content) => {
  return content.test(/jquery/);
}
```

注意：被忽略的模块不应该含有 import、require、define 的调用，或其他任何导入机制。因为浏览器环境下目前无法直接运行模块化代码。

14.4.2 rules

Rules 用于配置模块的读取和解析规则。这些规则能够修改模块的创建方式，对模块运用 Loader 或者修改解析器（Parser）。rules 是一个数组，数组的每一项都描述了如何处理符合规则的模块。

配置 rules 时一般需要完成以下工作：

- 匹配条件：通过 test、include、exclude 来指定要应用或排除 Loader 的文件。
- Loader 列表：对匹配成功的文件运用执行的 Loader，也可以传递数组，多个 Loader 的运用顺序是从后往前应用的。此外，每个 Loader 还支持传递选项。

下面通过一些示例来介绍具体的使用方法：

```
module: {
  rules: [ // rules 数组，每一项都是一个匹配规则
    {
      test: /\.css$/, // 匹配.css 扩展名的文件
      use: ['style-loader', 'css-loader'] // 运用 2 个 Loader，执行顺序为
      css-loader->style-loader
    },
    {
      test: /\.js$/, // 匹配.js 扩展名的文件
      use: ['babel-loader?cacheDirectory'],
      // 运用 1 个 Loader 并使用查询字符串形式传入选项
      include: path.resolve(__dirname, 'src'), // 只对 src 目录下的文件应用规则
    },
    {
      test: /\.less$/, // 匹配.less 扩展名文件
      use: ['style-loader','css-loader','less-loader'],
      exclude: path.resolve(__dirname, 'node_modules'),
      // 忽略 node_modules 目录中的文件
    },
    {
      test: /\.(gif|png|jpg)$/, // 匹配.gif 或.png 或.jpg 扩展名文件
      use: [ // 以对象方式运用 Loader
        {
          loader:'file-loader',
          options: { // 以对象方式传递选项
            name: 'images/[name].[ext]',
            esModule: false
          }
        }
      ]
    }
  ]
}
```

Webpack 以模块化的 JavaScript 为入口进行模块解析，内置有 AMD、CommonJS、ES6 等模块系统的支持。解析器可以对匹配的文件精确地控制哪些模块系统被解析、哪些模块系统不解析。

noParse 只能对文件级别进行控制，对于匹配的文件，任何模块系统都不会进行解析。下面针对 rules 匹配的文件设置模块系统解析规则。

```
{
 module: {
  rules: [
   {
    test:/\.js$/,
    use: ['babel-loader?cacheDirectory'],
    parser:{
      amd: false,         // 不解析 AMD
      commonjs: false,    // 不解析 commonjs
      harmony: false,     // 不解析 es6 import/export 模块语法
      requirejs: false,   // 不解析 requirejs
    }
   }
  ]
 }
}
```

14.5　Resolve

Webpack 内置 JavaScript 模块化语法解析功能，默认采用模块化标准中约定的规则去解析，通过 resolve 选项我们可以自定义 Webpack 的解析规则。

14.5.1　alias

创建 import 或 require 模块的别名，使得导入模块变得简单。例如，Vue 应用中常见的代码如下：

```
import Home from '@/Home.vue';
```

可以通过以下方式配置别名：

```
resolve: {
 alias: {
  @: './src'
 }
}
```

实际上导入的路径如下：

```
import Home from './src/Home.vue';
```

也可以在 alias 对象的键名后添加$号，以表示精确匹配：

```
resolve: {
  alias: {
    xyz$: path.resolve(__dirname, 'path/to/file.js'),
  }
}
```

这将产生以下结果：

```
import Test1 from 'xyz';          // 精确匹配，所以 path/to/file.js 被解析和导入
import Test2 from 'xyz/file.js';  // 非精确匹配，触发普通解析
```

14.5.2　extensions

当不带扩展名的模块导入时，Webpack 会读取本配置指定的扩展名去尝试访问文件是否存在。默认为：

```
extension: ['.js', '.json']
```

也就是说，当遇到 require('./home')这种不带扩展名的导入语句时，Webpack 会先寻找./home.js，如果该文件不存在，则继续寻找./home.json；如果所有扩展名都尝试之后仍无法找到文件，则报错提示模块不存在。

比如开发 Vue 应用时，可以使用以下配置，这样在导入 Vue 组件时就不用写扩展名了。

```
extensions:['.vue','.js','.json']
```

14.5.3　mainFields

当从 NPM 包中导入模块时，例如 import * as D3 from "d3"，该选项将决定在 package.json 中使用哪个字段导入模块。

根据 Webpack 配置中指定的目标模块不同，默认值也会不同。当目标模块的属性值设置为 webworker、web 或者未设置，默认值为：

```
mainFields: ["browser","module","main"]
```

对于其他的任意 target，默认值为：

```
mainFields: ["module", "main"]
```

例如，D3 的 package.json 含有这些字段：

```
{
  "main":"build/d3.Node.js",
  "browser":"build/d3.js",
  "module":"index"
}
```

这意味着当我们 import * as D3 from "d3"，实际上会先加载 browser 设置的文件 build/index.js。而如果我们打包 Node.js 环境下的程序时，默认会从 module 设置的文件 index 开始加载。

14.5.4　modules

该选项配置 Webpack 去哪些目录下寻找第三方模块，默认情况下只会去 node_modules 目录下寻找。如果项目中使用了其他模块，不在 node_modules 目录下时，就需要传递很长的相对路径，比如下面的目录结构：

```
|----lib
    |----picker
        |----index.js
        |----index.css
|----src
    |----pages
        |----home.js
```

home.js 导入 lib/picker/index.js 时，需要使用下面的代码：

```
import picker from '../../lib/picker'
```

如果在 modules 中配置 lib 目录，代码如下：

```
modules: [path.resolve(__dirname, 'lib'), 'node_modules']
```

home.js 的代码就可以简化了：

```
import picker from 'picker'
```

14.6　devtool

devtool 配置 Webpack 是否生成，以及如何生成 source-map。默认情况下 Webpack 不会生成 SourceMap，如果需要生成 SourceMap 文件以方便调试，可以使用以下配置：

```
devtool: 'source-map'
```

构建后的 JS 文件的同一级中会有相应的 map 文件。

14.7　externals

externals 用来告知 Webpack 应用代码中哪些模块是外部环境提供的，Webpack 构建时不要打包它们。

比如早期的 JavaScript 库都是通过全局变量来暴露变量的，例如著名的 jQuery 代码：

```html
<script src="js/jquery.min.js"></script>
```

JavaScript 文件加载成功后，会自动在全局空间下挂载$变量。通过$.xxx 的方式来调用 jQuery 的方法。

如果需要在模块化的环境下使用 jQuery，则需要使用下面的代码：

```js
import $ from 'jquery';

$('body').func();
```

执行 Webpack 构建后会发现构建结果中包含了 jQuery 相关的代码，而我们在 HTML 文件中已经引用过一份，这样会导致 jQuery 加载两次，无谓地浪费用户的网络资源。externals 就是针对该场景而生的。

以下是将 jQuery 声明为外部模块的配置：

```js
externals: {
  jquery: 'jQuery', // 键名代表应用代码中导入的模块名称，值代表全局变量
}
```

14.8 DevServer

在前面的内容中，我们介绍过可以提供 LiveReload 功能的 DevServer，本节我们来详细介绍一下。注意：DevServer 配置项只有 webpack-dev-server 模块，Webpack 本身不识别 DevServer 配置。

1. hot

是否启用模块热替换功能。DevServer 默认的 LiveReload 是检测到变更后刷新整个页面以此来实现实时预览。启用模块热替换功能后，只会替换变更的模块，这样在开发单页应用（Vue、React 等）时比较方便，LiveReload 刷新页面之后一些组件状态就丢失了。

2. contentBase

配置 DevServer HTTP 服务的 DocumentRoot。默认情况下，将使用当前的工作目录作为 DocumentRoot，但是也可以修改为其他值，比如，将 public 目录作为 DocumentRoot 提供服务。

```js
contentBase: path.join(__dirname, "public")
```

需要注意的是，DevServer 提供的 HTTP 服务器允许我们在浏览器中访问相关文件。DevServer 暴露的文件有两类：

- 本地文件，也就是项目目录中的文件。
- Webpack 构建结果，由于构建结果直接交付给了 DevServer，因此本地会看不到构建出的文件。

contentBase 只能配置本地文件的目录，如果不需要暴露本地文件到 HTTP 服务器上，可

以通过 contentBase: false 配置来关闭。

3. host

配置 DevServer 监听的地址。默认情况下为 127.0.0.1，只有本机能够访问。如果我们开发手机 H5 页面，那么在手机上预览时需要配置该选项为局域网 IP，才能在手机上访问 DevServer。

4. port

配置 DevServer 监听的端口，默认情况下为 8080。如果 8080 端口被占用，可以配置其他端口。

14.9 Plugins

Plugin 用于扩展 Webpack 的功能，Plugin 是通过 Webpack 在构建流程中回调钩子函数来实现的，因此几乎可以实现任何与构建相关的功能。

Plugin 的功能非常强大，但是配置起来却比较简单。plugins 配置项接收一个数组，数组中的元素为 Plugin 实例，同一个插件可以允许有几个实例同时存在。

```
plugins: [
  new HtmlWebpackPlugin({
    chunks: ['main'],
    filename: 'index.html',
    template: 'index.html'
  }),
  new HtmlWebpackPlugin({
    chunks: ['about'],
    filename: 'about.html',
    template: 'about.html'
  }),
]
```

每种插件支持的选项不尽相同，这个需要各位读者去熟悉一下插件的使用文档。

目前来说 Webpack 的生态非常强大，大部分场景下都有能够直接使用的插件来完成相应的功能。如果实在找不到插件，我们也可以根据需要自己来开发插件，这个将在后面的章节中进行介绍。

14.10 完整示例

在前面的章节中，我们详细介绍了常用配置项的具体含义，但是对于完整的项目来说，还没有一份完整的配置。下面结合一份完整的配置文件做一下说明：

```
module.exports = {
```

```
mode: 'production',          // 启用生产环境配置
entry: './app/entry',        // 项目入口文件
output: {
  path: path.resolve(__dirname, 'build'),              // 输出路径
  filename: '[name].[hash:16].js',
  publicPath: 'https://static.ddhigh.com/assets/',     // CDN 地址
  library: 'MyLibrary',      // 导出库的名称
  libraryTarget: 'umd',      // 导出库的类型
  chunkFilename: '[name].[chunkhash:16].js',           // chunk 文件的名称
},
module: {            // 模块配置
  rules: [           // Loader 配置
    {
      test: /\.js$/,
      include: [
        path.resolve(__dirname, 'src')
      ],
      exclude: [
        path.resolve(__dirname, 'lib')
      ],
      loader: 'babel-loader',
      options: {
        presets: ['es2015']
      }
    }
  ],
  noParse: [         // 不解析的模块配置
    /jquery/,
  ],
  resolve: {
    modules: [       // 用于查找模块的目录
      'node_modules',
      path.resolve(__dirname, 'lib'),
    ],
    extensions: ['.js','.json','.vue'],    // 模块扩展名
    alias: {         // 别名配置
      xyz$: path.resolve(__dirname, 'path/to/file.js'),
    },
    mainFields: ['main'],     // 模块的 package.json 中描述入口文件的属性名
  },
devtool: 'source-map',        // devtool 配置
context: __dirname,           // 上下文配置
devServer: {                  // 开发服务器配置
  hot: true,                  // HotReload, 模块热替换
  contentBase: path.join(__dirname, 'public'),
  // DevServer HTTP 服务器的文档根目录
  host: '127.0.0.1',          // DevServer 监听 IP
  port: 8080,                 // DevServer 监听端口
},
plugins: [                    // 插件配置
```

```
    ]
  }
};
```

14.11 本章小结

本章有关配置的相关知识看起来很多，实际上很多情况下我们使用默认的配置即可，没有必要都去记住它们，在理解配置项的情况下，根据具体需求来修改相关配置即可。

Loader 和 Plugin 是开发中用得比较多的，大部分情况下懂得使用社区提供的资源完成需求即可。

第15章

Vue 实战

本章内容

- 使用 Webpack 开发 Vue 应用
- Vue 应用添加 TypeScript 支持

Vue 是一个国人编写的、渐进式的 MVVM 框架,比 React 语法更灵活,比 Angular 上手门槛更低、更轻量。虽然 Vue 可以直接在浏览器中运行 Vue 的相关代码,但是为了方便编码以及提高工程效率,还是建议大家使用基于 Webpack 的工作流去开发 Vue 项目。

15.1 Hello World

Vue 和 React、Angular 的核心理念是一致的,都是通过数据驱动视图来进行开发,颠覆了传统应用中以操作 DOM 为核心的开发方式。

Vue 组件的 HTML、CSS 和 JavaScript 代码都写在一个以 .vue 为扩展名的文件中,非常方便组件化开发,下面以一个 Hello World 的例子进行说明。

项目目录如下:

```
|----
    |----src
        |---- App.vue # 根组件
        |---- main.js # 入口 JS
|---- index.html
|---- package.json
|---- webpack.config.js
|---- postcss.config.js
```

|---- .babelrc

先安装 Vue 模块：

```
npm install vue -save
```

index.html 代码：

```html
<!DOCTYPE html>
<html lang="en">
<head>
    <meta charset="UTF-8">
    <meta name="viewport" content="width=device-width, initial-scale=1.0">
    <title>Document</title>
</head>
<body>
    <div id="app"></div>
</body>
</html>
```

App.vue 代码：

```vue
<template>
  <div class="msg" @click="handleClick">{{msg}}</div>
</template>
<script>
export default {
  data() {
    return {
      msg: "hello world"
    };
  },
  methods: {
    handleClick() {
      console.log("clicked");
    }
  }
};
</script>
<style>
.msg {
  color: red;
}
</style>
```

main.js 代码：

```js
import Vue from 'vue';
import App from './App.vue';

new Vue({
    el: '#app',
```

```
render: h => h(App)
})
```

15.2 配置 Webpack

15.2.1 Loader 和 Plugin

通过 App.vue 文件中包含的 template、style 和 script 可知，至少需要以下 Loader：

- （必选）加载 .vue 文件的 Loader，需要将 template、style 和 script 三个标签下的代码交给相应的 Loader 处理，比如 style 交给 css-loader，等等。
- （必选）将 style 标签中的 CSS 编译为 JavaScript 代码的 Loader。
- （必选）将 template 标签中的代码编译为 JavaScript 代码的 Loader。
- （可选）script 标签中的代码使用 ES6 语法的 Loader。
- （可选）给 CSS 提供额外能力的 Loader。

通过以上分析，Vue 官方以及社区已经给出了答案：

- vue-loader：加载 .vue 文件，解析出 script、style 和 template 代码块中的代码，并把解析结果分别交给相应的 Loader 继续处理。
- css-loader+vue-style-loader：css-loader 用来将 vue-loader 提供的 CSS 代码转换为 JavaScript 代码，vue-style-loader 用来将转换后的 CSS 代码通过 DOM 操作插入到 HTML 页面的 style 中。
- vue-template-compiler：将 vue-loader 提取的 template 代码转换为 JavaScript 代码。
- babel-loader：给 JavaScript 提供 ES6 语法的能力。
- postcss-loader：给 CSS 提供额外能力。

此外，vue-loader 提供了 VueLoaderPlugin 插件，需要在构建时导入。

15.2.2 安装依赖模块

通过上文中的分析，我们需要安装以下依赖：

```
# 安装 babel 相关模块
npm install babel-loader @babel/core @babel/runtime @babel/preset-env
    @babel/plugin-transform-runtime
# 安装 vue 相关的 loader，会自动安装 vue-style-loader
npm install vue-loader vue-template-compiler
# 安装 css 相关模块
npm install postcss-loader css-loader autoprefixer
# 安装 html-webpack-plugin
npm install html-webpack-plugin
# 安装 webpack 相关模块
```

```
npm install webpack webpack-cli webpack-dev-server
```

15.2.3 编写配置文件

1. webpack.config.js

```js
const path = require('path');
const VueLoaderPlugin = require('vue-loader/lib/plugin');
const HtmlWebpackPlugin = require('html-webpack-plugin');

module.exports = {
    entry: './src/main',      // 入口配置
    output: {                  // 输出配置
        path: path.resolve(__dirname, 'build'),
        filename: '[name].[hash:8].js'
    },
    module: {
        rules: [
            {
                test: /\.vue$/, // 针对.vue扩展名处理
                loader: 'vue-loader'
            },
            {
                test: /\.css$/, // 针对css代码处理
                use: ['vue-style-loader', 'css-loader', 'postcss-loader']
            },
            {
                test: /\.js$/, // 针对js代码处理
                use: 'babel-loader'
            }
        ]
    },
    resolve: {
        extensions: ['.js', '.vue', '.json']
    },
    plugins: [
        new VueLoaderPlugin(),
        new HtmlWebpackPlugin({
            chunks: ['main'], // entry是字符串,因此chunk名称为main
            template: 'index.html',
            filename: 'index.html'
        })
    ]
};
```

2. .babelrc

```
{
    "plugins": [
        "@babel/plugin-transform-runtime"
    ],
```

```
    "presets": [
        "@babel/preset-env"
    ]
}
```

3. postcss.config.js

```
module.exports = {
    plugins: [
        require('autoprefixer')
    ]
};
```

4. package.json 添加 scripts：

```
{
    "scripts": {
        "start": "webpack-dev-server",
        "build": "webpack"
    }
}
```

15.2.4　执行构建

执行 npm run start 命令启动开发环境会自动启动 webpack-dev-server，访问 http://localhost:8080 即可看到应用。

15.3　生产构建

生产环境下 CSS 文件一般会提取到单独的文件，因此我们需要使用到 mini-css-extract-plugin 插件，这个插件在之前的章节中介绍过，这里不多说，我们直接来看相关配置。

15.3.1　Webpack 配置

在开发环境中使用 vue-style-loader 即可，但是在生产环境中需要使用 mini-css-extract-plugin 来提取 CSS。下面是 webpack.config.js 代码：

```
const path = require('path');
const VueLoaderPlugin = require('vue-loader/lib/plugin');
const HtmlWebpackPlugin = require('html-webpack-plugin');
const MiniCssExtractPlugin = require('mini-css-extract-plugin')

const config = {
    entry: './src/main',
    output: {
        path: path.resolve(__dirname, 'build'),
```

```
        filename: '[name].js'
    },
    module: {
        rules: [
            {
                test: /\.vue$/,
                loader: 'vue-loader'
            },
            {
                test: /\.css$/,
                use: [process.env.NODE_ENV === 'production' ?
    MiniCssExtractPlugin.loader : 'vue-style-loader', 'css-loader',
    'postcss-loader']
            },
            {
                test:/\.js$/,
                use: 'babel-loader'
            }
        ]
    },
    plugins: [
        new VueLoaderPlugin(),
        new HtmlWebpackPlugin({
            chunks: ['main'],
            template: 'index.html',
            filename: 'index.html'
        })
    ]
};
if (process.env.NODE_ENV === 'production') { // 在生产环境中添加 css 提取插件
    config.plugins.push(new MiniCssExtractPlugin({
        filename: '[name].css'
    }))
}
module.exports = config;
```

15.3.2　package.json 修改

在配置文件中可以使用环境变量来区分是开发环境还是生产环境,因此我们需要编辑 package.json 的 scripts:

```
{
  "scripts": {
    "start": "NODE_ENV=development webpack-dev-server",
    "build": "NODE_ENV=production webpack"
  }
}
```

15.4 TypeScript 支持

新版本的 Vue（2.5.0+）提供了对 TypeScript 的支持。相信使用 TypeScript 来编写 JavaScript 应用会越来越流行。下面讲解如何在刚才项目的基础上接入 TypeScript 支持。

TypeScript 文件的扩展名是 .ts，Webpack 无法直接使用，因此需要 ts-loader。此外，ts-loader 需要配合 typescript 模块工作，因此需要安装以下依赖：

```
npm install typescript ts-loader --save-dev
```

15.4.1 TypeScript 配置

TypeScript 的编译器 tsc 需要读取项目根目录下的 tsconfig.json 配置，新建 tsconfig.json，代码如下：

```
{
  "compilerOptions": {
    // 与 Vue 的浏览器支持保持一致
    "target": "es5",
    // 这可以对 this 上的数据属性进行更严格的推断
    "strict": true,
    // 如果使用 webpack 2+ 或 rollup，可以利用 tree-shake:
    "module": "es2015",
    "moduleResolution": "node"
  },
  "include":[
    "src/**/*.vue",
    "src/**/*.ts"
  ]
}
```

此外，由于 TypeScript 不识别 .vue 扩展名这种文件，因此我们可以通过声明 module 的方式来解决：

```
src/vue-shim.d.ts

declare module '*.vue' {
    import Vue from 'vue';
    export default Vue;
}
```

这样，TypeScript 在导入 .vue 文件时就知道这个是 Vue 组件了。

15.4.2 Webpack 配置

Webpack 的配置改动比较小，主要是以下几个方面：

- 添加 ts-loader 处理 .ts 文件。
- 由于项目不使用 JavaScript 编写，因此可以移除 babel-loader 相关配置。
- resolve.extensions 添加 .ts，导入 ts 代码时不用添加扩展名。

支持 TypeScript 的 webpack.config.js 如下：

```
const path = require('path');
const VueLoaderPlugin = require('vue-loader/lib/plugin');
const HtmlWebpackPlugin = require('html-webpack-plugin');
const MiniCssExtractPlugin = require('mini-css-extract-plugin')

const config = {
    entry: './src/main',
    output: {
        path: path.resolve(__dirname, 'build'),
        filename: '[name].js'
    },
    module: {
        rules: [
            {
                test: /\.vue$/,
                loader: 'vue-loader'
            },
            {
                test: /\.css$/,
                use: [process.env.NODE_ENV === 'production' ?
MiniCssExtractPlugin.loader : 'vue-style-loader', 'css-loader',
'postcss-loader']
            },
            {
                test: /\.ts$/,
                loader: 'ts-loader',
                exclude: /node_modules/,
                options: { appendTsSuffixTo: [/\.vue$/] }
            }
        ]
    },
    resolve: {
        extensions: ['.ts', '.js', '.vue', '.json']
    },
    plugins: [
        new VueLoaderPlugin(),
        new HtmlWebpackPlugin({
            chunks: ['main'],
```

```
            template: 'index.html',
            filename: 'index.html'
        })
    ]
};

if (process.env.NODE_ENV === 'production') {
    config.plugins.push(new MiniCssExtractPlugin({
        filename: '[name].[hash:8].css'
    }))
}
module.exports = config;
```

15.4.3　App.vue

App.vue 的 script 代码需要改为 ts，为了演示 ES6 中的 async/await 语法，给 App.vue 添加了点击事件，并异步调用了一个 sum 方法。

```
<template>
  <div class="msg" @click="handleClick">{{msg}}</div>
</template>
<script lang="ts">
import Vue from "vue";
export default Vue.extend({
  data() {
    return {
      msg: "hello world"
    };
  },
  methods: {
    async handleClick() {
      const data = await this.sum(1, 2);
      console.log(data);
    },
    sum(a: number, b: number) {
      return new Promise(resolve => {
        setTimeout(() => {
          resolve(a + b);
        });
      });
    }
  }
});
</script>
<style>
.msg {
  color: red;
}
</style>
```

script 标签需要添加 ts，此外需要导入 vue 模块，并将导出的对象通过 Vue.extend 函数进行处理。

15.5　本章小结

本章从零开始搭建了基于 JavaScript 和 TypeScript 语言的 Vue 项目，使用 Webpack 作为构建工具完整地构建了一个 Hello World 项目，方便各位读者学习和调试。

基于 Webpack 的项目开发中一般都是主要配置 Loader 和 Plugin，有些 Loader 可能有独立配置或依赖，比如 babel-loader 和 ts-loader，此时根据文档进行安装和配置即可。

第 16 章

React 实战

本章内容

- Babel 与 JSX 语法介绍
- 使用 Babel 开发 React 项目
- 使用 TypeScript 开发 React 项目

React 是一个用于构建用户界面的 JavaScript 库。React 由 Facebook 开发，在 2013 年 5 月开源，经过数年的发展，已经拥有相当完整的生态系统和工作流，使得复杂应用的开发变得相当便利。

16.1　JSX

React 使用了名为 JSX 的语法来编写组件，例如：

```
class AppComponent extends React.Component {
  render() {
    return (
      <div onClick={this.onClick}>Hello World</div>
    );
  }

  // 点击处理
  onClick = () => {
    console.log('clicked');
  };
}
```

JavaScript 代码用来编写业务逻辑，JSX 用来渲染视图，这是 React 组件的一般写法。但是 JSX 语法目前没有任何 JavaScript 引擎能直接运行，需要借助代码编译器将 JSX 代码转换为下列的标准 JavaScript 代码：

```
React.createElement('div', null, 'Hello World');
```

目前 Babel 和 TypeScript 都提供了对 JSX 语法的支持和编译。下面来看一下如何使用 Webpack 和 Babel 来编写 React 应用。

16.2 Babel

Babel 的基础知识在前面的内容中已经介绍过，这里不再赘述。Babel 通过 @babel/preset-react 包来构建 React 应用。下面我们来手动搭建一个项目。

新建一个文件夹取名为 react-babel-example，进入该文件夹执行命令：

```
npm init -y # 初始化 package.json
npm install webpack webpack-cli webpack-dev-server --save-dev # 安装 Webpack

# 安装 Babel
npm install babel-loader @babel/core @babel/runtime
      @babel/plugin-transform-runtime @babel/preset-react @babel/preset-env -D
npm install html-webpack-plugin
npm install react react-dom # 安装 React 相关包
```

react-dom 以前是和 React 在一起的，由于 React Native 项目的原因，react-dom 被拆分出来用来将 React 应用渲染到 Web 环境下，同理 React Native 可以将 React 应用渲染到 Native APP 环境下。

1. 目录结构

目录结构：

```
|----src
   |----App.jsx # 主组件
   |----main.js # 入口文件
|----.babelrc
|----webpack.config.js
|----index.html
|----package.json
```

- index.html 文件和 Babel 下一致。
- package.json 的 build、start 命令与 Babel 一致。

webpack.config.js

```
const path = require('path');
const HtmlWebpackPlugin = require('html-webpack-plugin');
```

```js
module.exports = {
    entry: './src/main',      // 入口配置
    output: {
        path: path.resolve(__dirname, 'build'),
        filename: '[name].[hash:8].js'
    },
    module: {
        rules: [
            {
                test: /\.jsx?$/, // 配置以 jsx 和 js 为后缀的文件由 Babel 处理
                use: 'babel-loader'
            }
        ]
    },
    devServer: {
        hot: true,             // 启用热加载
    },
    plugins: [
        new HtmlWebpackPlugin({
            chunks: ['main'],
            template: 'index.html',
            filename: 'index.html'
        })
    ],
    resolve: {
        extensions: ['.js', '.jsx', '.json'] // 添加 jsx 后缀，导入组件时不用添加 .jsx
    }
};
```

.babelrc

```json
{
    "plugins": [
        "@babel/transform-runtime"
    ],
    "presets": [
        "@babel/react",
        "@babel/env"
    ]
}
```

index.html

```html
<!DOCTYPE html>
<html lang="en">
<head>
    <meta charset="UTF-8">
    <meta name="viewport" content="width=device-width, initial-scale=1.0">
    <title>Hello World</title>
</head>
<body>
    <div id="app"></div>
```

```
</body>
</html>
```

id 为 app 的 div 用来作为 React 渲染根节点。

src/App.jsx

```
import React from 'react';

export default class App extends React.Component {
    render() {
        return (
            <div onClick={this.handleClick.bind(this)}>Hello World</div>
        );
    }

    handleClick() {
        alert('Hello');
    }
}
```

src/main.js

```
import React from 'react';
import ReactDOM from 'react-dom';
import App from './App';

ReactDOM.render(<App />, document.querySelector('#app'));
```

使用 JSX 语法的文件都必须导入 react 包。

package.json

package.json 添加了开发编译和生产编译两条命令：

```
{
    "scripts": {
        "start": "webpack-dev-server",
        "build": "webpack"
    }
}
```

2. 执行构建

执行 npm run start 开启开发服务器，访问 http://localhost:8080 即可看到"Hello World"。

16.3　TypeScript

TypeScript 原生支持 JSX 语法，不需要安装额外的编译器，不过由于早期的 React 不是 TypeScript 编写的，因此在 TypeScript 中使用 React，需要导入 react 和 react-dom 的声明文件才可以使用。

下面我们使用 Webpack+TypeScript 搭建一个 React 应用。

新建一个文件夹取名为 react-typescript-example，进入该文件夹执行命令：

```
npm init -y
npm install webpack webpack-cli webpack-dev-server --save-dev
npm install typescript ts-loader      # 安装 TypeScript 相关包
npm install react react-dom --save  # 安装 React

# 安装 React 的 TypeScript 声明文件
npm install @types/react @types/react-dom --save-dev
npm install html-webpack-plugin
```

1. 目录结构

```
|----src
    |----App.tsx
    |----main.tsx
|----package.json
|----tsconfig.json
|----webpack.config.js
```

tsconfig.json

tsconfig.json 是 TypeScript 编辑器 tsc 使用的配置文件，内容如下：

```
{
  "compilerOptions": {
    "target": "es5",
    "module": "commonjs",
    "jsx": "react",          // JSX 语法配置
    "strict": true,
    "lib": [
      "DOM"
    ],
    "esModuleInterop": true
  },
  "include": [
    "./src/**/*.ts",
    "./src/**/*.tsx"
  ],
  "exclude": [
    "node_modules"
  ]
}
```

webpack.config.js

```
const path = require('path');
const HtmlWebpackPlugin = require('html-webpack-plugin');

module.exports = {
    entry: './src/main',
    output: {
```

```js
            path: path.resolve(__dirname, 'build'),
            filename: '[name].[hash:8].js'
        },
        module: {
            rules: [
                {
                    test: /\.tsx?$/,        // tsx 和 ts 文件使用 ts-loader 加载
                    use: 'ts-loader'
                }
            ]
        },
        plugins: [
            new HtmlWebpackPlugin({
                chunks: ['main'],
                filename: 'index.html',
                template: 'index.html'
            })
        ],
        resolve: {
            extensions: ['.ts', '.tsx', '.js', '.json']    // 添加 ts 和 tsx 后缀
        }
};
```

src/App.tsx

```tsx
import React from 'react';

export default class App extends React.Component {
    render() {
        return (
            <div>Hello World</div>
        )
    }
}
```

src/main.tsx

```tsx
import React from 'react';
import ReactDOM from 'react-dom';
import App from './App';

ReactDOM.render(<App />, document.querySelector('#app'));
```

2. 执行构建

执行 npm run start 开启开发服务器，访问 http://localhost:8080 即可看到"Hello World"。

16.4　本章小结

本章从零开始搭建了基于 Babel 和 TypeScript 语言的 React 项目，使用 Webpack 作为构建工具完整地构建了一个 Hello World 项目，方便各位读者学习和调试。

实际开发中 React 官方提供了 Babel 和 TypeScript 的脚手架，但是 Webpack 相关的脚本是隐藏的，为了解决读者在使用这些脚手架时可能遇到的问题，本章通过一步步添加的形式对 Webpack 和 React 的工作机制进行了介绍，方便各位读者了解原理。

第 17 章

服务端渲染

本章内容

- 服务端渲染原理
- React 的服务端渲染实战

使用 Vue/React/Angular 开发的单页应用（只有一个入口 index.html）都是通过 JavaScript 在浏览器中渲染出来的，这种方式主要存在两个比较明显的缺点：

- 大部分搜索引擎无法收录我们的网页，因为展示的界面都是由 JavaScript 异步渲染的。
- 浏览器渲染在性能比较差的移动端设备上，可能导致首次加载速度比较慢，影响体验。

为了解决这两个问题，社区出现了服务端渲染（SSR，Server Side Rendering）的解决方案，可以让 JavaScript 在服务端渲染完生成的 HTML 之后再返回给客户端，从而解决以上两个问题。

到目前为止，主流的 Vue/React/Angular2 都支持服务端渲染，其中最早支持的是 React，本章以 React 为例来介绍如何用一份代码通过 Webpack 构建浏览器和服务器环境的应用（也称同构应用）。

React 提出的虚拟 DOM 让 React 与渲染层实现了分离，从而有了 react-dom 和 react-native 等渲染层实现。而在服务端来说，只需要将虚拟 DOM 渲染为 HTML 字符串即可。

同构应用的核心目的是：使同一份应用代码分别编译为浏览器和服务器环境下的代码。服务器环境使用 Node.js，因此有以下注意事项：

- 不能使用浏览器环境提供的 API，比如 DOM 等，这些 API 在 Node.js 下不可用。
- 不能将第三方 node_modules 模块打包，而需要通过 CommonJS 规范的 require 函数进行加载。

接下来我们使用上一章写过的 React TypeScript 进行 SSR 同构渲染。

17.1　SSR 原理

（1）使用 react-dom 提供的 renderToString 函数将 React 应用导出为一个 Node.js 模块，该模块导出一个 render 函数。

（2）使用 express 开启 HTTP 服务器接收请求，在响应中将第 1 步的函数渲染输出到客户端。

简要步骤如下：

（1）添加 SSR 下的应用入口文件。
（2）添加 SSR 环境下打包使用的 Webpack 配置文件。
（3）使用 HTTP 服务器承载 HTTP 请求以输出 HTML。

17.2　添加 SSR 的 webpack.config.js

在项目根目录新建 webpack.ssr.config.js，内容如下：

```js
const path = require('path');

module.exports = {
    entry: './src/main.ssr',       // 入口文件
    target:'node',                 // 导出为 Node.js 环境
    output: {
        libraryTarget:'commonjs2', // 打包为 commojs2 模块，可以被 Npde.js 直接加载
        path: path.resolve(__dirname, 'build'),
        filename: 'ssr.bundle.js'
    },
    module: {
        rules: [
            {
                test: /\.tsx?$/,
                use: 'ts-loader'
            }
        ]
    },
    resolve: {
        extensions: ['.ts', '.tsx', '.js', '.json']
    }
};
```

原来的 webpack.config.js 修改如下：

```js
const path = require('path');
const HtmlWcbpackPlugin = require('html-webpack-plugin');

module.exports = {
    entry: './src/main',
    output: {
        path: path.resolve(__dirname, 'build'),
        filename: 'bundle.js' // 输出文件名变更
    },
    module: {
        rules: [
            {
                test: /\.tsx?$/,
                use: 'ts-loader'
            }
        ]
    },
    plugins: [
        new HtmlWebpackPlugin({
            chunks: ['main'],
            filename: 'index.html',
            template: 'index.html'
        })
    ],
    resolve: {
        extensions: ['.ts', '.tsx', '.js', '.json']
    }
};
```

Node.js 下的构建脚本变更主要有以下几点：

- 入口文件变更。
- target 修改为 node，打包时不会打包 Node.js 内置的模块打包。
- libraryTarget 修改为 commonjs2，以便被 Node.js 加载。
- 去掉 HtmlWebpackPlugin 插件。

17.3 添加 SSR 的入口文件

新建 src/main.ssr.tsx，内容如下：

```tsx
import React from 'react';
import { renderToString } from 'react-dom/server';
import App from './App';

export function render() {
    return renderToString(<App />)
}
```

17.4 添加 SSR 打包命令

package.json 需要添加 build-ssr 命令用来打包 SSR 环境下的 bundle，内容如下：

```
{
  "scripts": {
    "build": "webpack",
    "start": "webpack-dev-server",
    "build-ssr":"webpack --config webpack.ssr.config.js" # 新增
  }
}
```

17.5 执行构建

执行以下命令进行构建：

```
npm run build          # 打包浏览器环境代码
npm run build-ssr      # 打包 SSR 环境代码
```

17.6 添加 Node.js HTTP 服务器

HTTP 服务器使用 express 进行开发，步骤如下：

（1）安装 express 依赖：

```
npm install express --save
```

（2）项目根目录添加 ssr.js：

```
const express = require('express');                        // 加载 express
const { render } = require('./build/ssr.bundle');          // 加载 ssr

const app = express();
app.get('/', (req, res) => {                               // 所有请求都到这里处理
    res.send(`
    <html>
    <head>
        <meta charset="UTF-8">
        <title>SSR</title>
    </head>
    <body>
```

```
            <div id="app">${render()}</div>
            <script src="./build/bundle.js"></script>
        </body>
        </html>
        `)
});
app.use(express.static("."))      // 将当前目录作为 HTTP 服务器文档根目录
app.listen(3000,()=>console.log('app listen on 3000'));
```

17.7　目录结构

构建完毕的目录结构如下:

```
|----build
    |----bundle.js              # 浏览器端的 budle.js
    |----index.html             # 入口文件
    |----ssr.bundle.js          # SSR bundle.
|----node_modules
|----src
    |----App.tsx                # 应用根组件
    |----main.ssr.tsx           # SSR 入口文件
    |----main.tsx               # 浏览器环境入口文件
|----index.html                 # 浏览器环境 index.html
|----package.json
|----ssr.js                     # express 入口 js
|----tsconfig.json
|----webpack.config.js          # 浏览器环境构建配置
|----webpack.ssr.config.js      # SSR 环境构建配置
```

17.8　运行应用

运行应用之前请确认执行了 npm run build-ssr 命令,否则 build 下将没有 bundle.ssr.js 文件。

执行以下命令启动 express:

```
node ssr.js
```

启动成功后终端将输出:

```
app listen on 3000
```

此时打开浏览器访问 http://localhost:3000 可以看到输出成功,页面源代码如下:

```
<html>
    <head>
```

```
        <meta charset="UTF-8">
        <title>SSR</title>
    </head>
    <body>
        <div id="app"><div data-reactroot="">Hello World</div></div>
        <script src="./build/bundle.js"></script>
    </body>
</html>
```

17.9 本章小结

在本章中,我们使用 Webpack 通过同一份应用代码构建了浏览器和服务端分别使用的应用文件,可以比较方便地解决本章开始时提出的问题,Vue/Angular 框架的服务端渲染步骤也是类似的,不过这些框架官方已经提供了比较完善的解决方案。本章从原理入手,一步一步搭建了 React 下的 SSR 应用,以方便读者理解服务端的渲染。

第 18 章

多页应用脚手架

本章内容

- 使用 Webpack 构建传统多页网站

在前面的实战中都是以 SPA（Single Page Application，单页应用）框架进行开发的，但是大部分的网站还是使用 HTML+CSS+JavaScript 的开发方式。传统网站一般不使用构建工具，或者使用 Grunt 和 Gulp 等非模块化构建工具，实际上这些网站在开发过程中会遇到以下问题：

（1）公共 CSS/JavaScript 如何管理？依赖关系如何处理？
（2）如何将 CSS/图片/JavaScript 部署到 CDN？
（3）浏览器兼容性问题？（主要指 CSS）
（4）如何使用 JavaScript 最新特性，如 ES6？
（5）多人协作时如何开发？
（6）图片资源如何方便地压缩？

实际上，对于以上问题，Webpack 都可以很好地解决，现在我们一起来学习一下。

18.1 项目结构

针对本章开头提出的问题，我们使用以下工具进行解决：

- 模块化/CDN 通过 Webpack 处理。
- 浏览器兼容性问题通过 PostCSS 处理。
- JavaScript 新语法问题通过 Babel 处理。

- 图片通过 image-webpack-loader 进行压缩。

传统企业网站或者新闻网站一般都是由首页、新闻页、关于页等页面组成的，因此我们需要将公共的 CSS 和 JavaScript 提取出来，另外每个网页会有自己的业务逻辑和样式，因此我们可以用以下结构作为我们的脚手架：

```
|----build 构建结果
|----src 源码目录
     |---- common 公共代码
          |----scss 公共 SCSS
          |----js 公共 JS
     |---- images 图片
     |---- pages 页面列表
          |----home 主页
               |---- index.js 主页 JS 逻辑
               |---- index.scss 主页 SCSS
          |----news 新闻页
               |---- index.js 新闻页 JS
               |---- index.scss 主页 SCSS
          |----about 关于页
               |---- index.js 关于页 JS
               |---- index.scss 关于页 SCSS
|----package.json
|----webpack.config.js    # webpack 配置
|----.babelrc             # Babel 配置
|----postcss.config.js    # PostCSS 配置
```

18.2 开发步骤

18.2.1 初始化项目与安装依赖

1. 初始化项目

```
mkdir multi-page-app
cd multi-page-app
npm init -y
```

2. 安装依赖

```
npm install webpack webpack-cli webpack-dev-server --save-dev        # Webpack
npm install babel-loader @babel/core @babel/runtime @babel/preset-env
    @babel/plugin-transform-runtime --save-dev                       # Babel
npm install sass-loader css-loader postcss-loader autoprefixer
    mini-css-extract-plugin node-sass autoprefixer --save-dev        # CSS 相关模块
npm install html-webpack-plugin html-loader --save-dev               # HTML 相关模块
npm install file-loader image-webpack-loader --save-dev              # 图片相关模块
npm install clean-webpack-plugin --save-dev                          # 清空 build 目录插件
```

html-loader 用来处理 HTML 页面中导入的资源。

18.2.2 配置

1. Webpack 配置

```
const path = require('path');
const MiniCssExtractPlugin = require('mini-css-extract-plugin');
const HtmlWebpackPlugin = require('html-webpack-plugin');
const { CleanWebpackPlugin } = require('clean-webpack-plugin');

module.exports = {
    entry: {       // 入口配置，每个页面一个入口
        home: './src/pages/home',
        news: './src/pages/news'
    },
    output: {      // 输出配置
        path: path.resolve(__dirname, 'build'),
        filename: 'js/[name].[hash:8].js',
        publicPath: '/'
    },
    module: {
        rules: [
            {
                test: /\.js$/,
                use: 'babel-loader',
                exclude: [
                    /node_modules/
                ]
            },
            {
                test: /\.scss$/,
                use: [MiniCssExtractPlugin.loader, 'css-loader', 'postcss-loader',
 'sass-loader']
            },
            {
                test: /\.css$/,
                use: [MiniCssExtractPlugin.loader, 'css-loader', 'postcss-loader']
            },
            {
                test: /\.(png|svg|jpg|gif)$/, // 图片类文件通过 file-loader 加载
                use: [
                    {
                        loader: 'file-loader',
                        options: {
                            esModule: false,
                            name: 'images/[name].[hash:8].[ext]'
                        }
                    },
                    {
```

```
                    loader: 'image-webpack-loader',  // 图片压缩 loader
                }
            ]
        },
        {
            test: /\.html$/,              // 处理 HTML 中导入的资源（如图片）
            use: 'html-loader'
        }
    ]
},
plugins: [
    new CleanWebpackPlugin(),      // 清空 build 目录，读取 output.path 配置
    new HtmlWebpackPlugin({         // 一个页面一个插件实例
        chunks: ['home'],
        template: './src/pages/home/index.html',
        filename: 'index.html'
    }),
    new HtmlWebpackPlugin({
        chunks: ['news'],
        template: './src/pages/news/index.html',
        filename: 'news.html'
    }),
    new MiniCssExtractPlugin({
        filename: 'styles/[name].[contenthash:8].css',
        chunkFilename: 'styles/[name].[contenthash:8].css'
    }),
]
};
```

2. Babel 配置

```
{
    "plugins": [
        "@babel/transform-runtime"
    ],
    "presets": [
        "@babel/env"
    ]
}
```

3. PostCSS 配置

```
module.exports = {
    plugins: [
        require('autoprefixer')
    ]
}
```

至此工作流部分就配置完成了，我们接下来进行业务代码的编写。

18.3 业务代码

由于篇幅限制,这里只列出部分业务代码,其他代码可以前往 GitHub 查看。

1. src/pages/home/index.html

```html
<!DOCTYPE html>
<html lang="en">

<head>
    <meta charset="UTF-8">
    <meta name="viewport" content="width=device-width, initial-scale=1.0">
    <title>Home</title>
</head>

<body>
    <header class="navbar">
        <nav>
            <ul>
                <li class="active"><a href="/">Home</a></li>
                <li><a href="/news.html">News</a></li>
            </ul>
        </nav>
    </header>

    <h1 class="title">Home</h1>
    <img src="../../images/main.jpg" alt="">
</body>

</html>
```

2. src/pages/home/index.js

```js
import './index.scss';

console.log('index loaded');
```

3. src/pages/home/index.scss

```scss
@import '../../common/css/common.scss';

.title {
    color: yellow;
}
```

4. src/pages/news/index.html

```html
<!DOCTYPE html>
<html lang="en">
```

```html
<head>
    <meta charset="UTF-8">
    <meta name="viewport" content="width=device-width, initial-scale=1.0">
    <title>News</title>
</head>

<body>
    <header class="navbar">
        <nav>
            <ul>
                <li><a href="/">Home</a></li>
                <li class="active"><a href="/news.html">News</a></li>
            </ul>
        </nav>
    </header>
    <h1 class="title">News</h1>
</body>

</html>
```

5. src/pages/news/index.js

```
import './index.scss';

console.log('news loaded');
src/pages/news/index.scss
@import '../../common/css/common.scss';

.title {
    color: red;
}
```

主页和新闻页都使用了 common.scss 中的导航栏样式代码，由此实现了代码复用。JavaScript 需要复用的代码也可以放到 src/common/js 目录中，页面 JS 通过 import 进行导入。

注意：本章完整示例代码地址为 https://github.com/nodejs-inaction/webpack-multipage-boilerplate。

18.4 本章小结

实战部分列举了 Webpack 常用的 Loader 和 Plugin，介绍了用来开发 SPA 应用或者传统应用的方法。不过 Webpack 还有更多的用法等待各位读者去挖掘，总体思路都是一致的，合理利用 Loader 和 Plugin 基本可以满足大部分场景下的需求。

接下来我们将学习如何对 Webpack 进行优化，比如减少输出大小，提高编译速度，等等。

第 19 章

性能优化

本章内容

- 优化 Webpack 构建速度与输出大小

19.1 限定 Webpack 处理文件范围

项目比较小的情况下 Webpack 的性能问题几乎可以忽略，但是一旦项目复杂度上升，Webpack 会有额外的性能损失，需要我们进行优化。

通过前面章节的学习，我们知道 Webpack 主要完成下面这些事情：

（1）通过 entry 指定的入口脚本进行依赖解析。
（2）找到文件后通过配置的 loader 对其进行处理。

因此，我们可以从这方面入手进行优化，减少 Webpack 搜索文件的范围，减少不必要的处理。

1. loader 配置

在之前的章节中介绍过 Loader 可以使用 test、include 和 exclude 配置项来匹配需要 Loader 处理的文件，因此推荐每个 Loader 在定义 test 之后还需要定义 include 或 exclude。

```
module.exports = {
  module:{
    rules:[
      {
        test:/\.js$/,
```

```
      use:'babel-loader',
      include: path.resolve(__dirname, 'src'), // 只处理src目录下的js文件
    }
  ]
 }
};
```

2. resolve.extensions 配置

导入未添加扩展名的模块时，Webpack 会通过 resolve.extensions 后缀去检查文件是否存在。由于 resolve.extensions 是一个数组，如果数组项比较多，正确的后缀放置得越靠后，Webpack 尝试次数就会越多，因而会影响到性能，因此配置 resolve.extensions 时需要遵守以下规则：

- 尽量减少后缀列表，不要将不存在的文件后缀配置进来。
- 出现频率越高的后缀尽量写到前面，比如可以将 .js 写在第一个。
- 业务代码中导入模块时，可以手动加上后缀导入，省去 Webpack 的查找过程。

3. module.noParse 配置

module.noParse 可以告诉 Webpack 忽略未采用模块系统文件的处理，可以有效地提高性能。比如常见的 jQuery 非常大，又没有采用模块系统，让 Webpack 解析这种类型的文件完全是浪费性能。

因此我们可以配置如下的 module.noParse：

```
module.exports = {
 module:{
   noParse:[/jQuery/]
 }
};
```

4. IgnorePlugin

在导入模块时，IgnorePlugin 可以忽略指定模块的生成。比如 moment.js 在导入时会自动导入本地化文件，一般情况下它几乎不使用而且又比较大，此时可以通过 IgnorePlugin 忽略对本地化文件的生成，以减小文件大小。

```
module.exports = {
 plugins:[
   new webpack.IgnorePlugin(/\.\/local/, /moment/)
 ]
};
```

19.2　DllPlugin

使用过 Windows 操作系统的读者应该会经常看到以 .dll 扩展名的文件，这些文件叫作动态链接库，包含了其他程序或动态链接库的函数和数据。

Webpack 的 DllPlugin 的思想是类似的，先将公共模块打包为独立的 Dll 模块，然后在业

务代码中直接引用这些模块。采用 DllPlugin 之后会大大提升 Webpack 构建的速度，原因在于，包含大量复用模块的动态链接库只需编译一次，之后的构建中会直接引用这些构建好的模块。

在 Webpack 中使用动态链接库有以下两个步骤：

- 通过 webpack.DllPlugin 插件打包出 Dll 库。
- 通过 webpack.DllReferencePlugin 引用打包好的 Dll 库。

下面以 React 项目为例进行说明。

Dll 库需要单独构建，因此我们需要一份单独的配置 Webpack 文件。

（1）新建 webpack.dll.config.js：

```javascript
const webpack = require('webpack');

module.exports = {
  entry:{
    react: ['react', 'react-dom']
  },
  output:{
    filename: '_dll_[name].js',              // 输出的文件名
    path: path.resolve(__dirname, 'dist'),   // 输出到dist目录
    library: '_dll_[name]'
  },
  plugins: [
    // name 要等于 output.library 里的 name
    new webpack.DllPlugin({
      name: "_dll_[name]",
      path: path.resolve(__dirname, "dist", "manifest.json") // 清单文件的路径
    })
  ]
};
```

（2）编辑 webpack.config.js：

```javascript
const webpack = require('webpack');

module.exports = {
  entry: './src/main',
  output:{
    filename: '[name].js',                   // 输出的文件名
    path: path.resolve(__dirname, 'dist'),   // 输出到dist目录
  },
  plugins: [
    // 传入 manifest.json
    new webpack.DllReferencePlugin({
      manifest: path.resolve(__dirname, "dist", "manifest.json") // 清单文件的路径
    })
  ]
};
```

（3）添加构建命令：

```
{
  "scripts":{
    "build-dll":"webpack --config webpack.dll.config.js",
    "build":"webpack"
  }
}
```

（4）构建 Dll：

```
npm run build-dll
```

（5）构建应用：

```
npm run build
```

Dll 需要先构建，否则应用将构建失败。

19.3　HappyPack

Webpack 默认情况下是单进程执行的，因此无法利用多核优势，通过 HappyPack 可以变成多进程构建，从而提升构建的速度。下面我们一起来看看如何使用 HappyPack 来加速构建。

（1）安装 HappyPack：

```
npm install happypack
```

（2）编辑配置文件，需要将 Loader 配置到 HappyPack 插件中，由 HappyPack 对 Loader 进行调用。

```
const HappyPackPlugin = require('happypack');
const path = require('path');

module.exports = {
  entry: './src/main',
  output:{
    path: path.resolve(__dirname, 'build'),
    filename:'[name].js'
  },
  module:{
    rules:[
      {
        test:/\.js$/,
        use:'happypack/loader?id=js',        // 配置 id 为 js
        include:[
          path.resolve(__dirname,'src')
        ]
      },
      {
```

```
        test:/\.scss$/,
        use:'happypack/loader?id=scss',     // 配置id为scss
        include:[
          path.resolve(__dirname,'src')
        ]
      },
      {
        test:/\.css$/,
        use:'happypack/loader?id=css',      // 配置id为css
        include:[
          path.resolve(__dirname,'src')
        ]
      }
    ]
  },
  plugins:[
    new HappyPackPlugin({
      id:'js',           // id为js的loader配置
      use:[
        {
          loader:'babel-loader',
          options:{
            plugins:['@babel/transform-runtime'],
            presets:['@babel/env']
          }
        }
      ]
    }),
    new HappyPackPlugin({
      id:'scss',         // id为scss的loader配置
      use:['style-loader','css-loader','sass-loader']
    }),
    new HappyPackPlugin({
      id:'css',          // id为css的loader配置
      use:['style-loader','css-loader']
    }),
  ]
};
```

19.4　Tree-Shaking

Tree-Shaking 原始的本意是"摇动树",这样就会将一些分支"摇掉",从而减少主干的大小。而 Webpack 中的 Tree-Shaking 是类似的,在 Webpack 项目中,有一个入口文件,相当于树的主干,入口文件又依赖了许多模块。在实际开发中,虽然依赖了某个模块,但其实只使用了其中的部分代码,通过 Tree-Shaking,可以将模块中未使用的代码剔除掉,从而减少构建结果的大小。

注意：只有使用 ES6 模块系统的代码，在 mode 为 production 时，Tree-Shaking 才会生效。因此，在编写代码时尽量使用 import/export 的方式。

19.5 按需加载

在开发中，我们一般会将业务代码打包为 app.js，其他第三方依赖打包为 vendor.js。这样会有一个比较大的问题，如果依赖的第三方模块过多，vendor.js 会越来越大，而在浏览器加载时需要完全加载完 vendor.js 才可以，这样就会造成无谓的等待，因为我们当前页面可能只使用了一部分代码。此时可以使用 Webpack 来实现按需加载，只有在真正用到这个模块时才会加载相应的 JS。

比如基于 echarts 开发了一个数据可视化页面，可以在这个路由组件下面使用异步的方式加载 echarts 的代码：

```
import('echarts').then(modules => {
  const echarts = modules.default;
  const chart = echarts.init(document.querySelector('#chart'));
});
```

不过使用按需加载时，构建代码中会包含 Promise 调用，因此低版本浏览器需要注入 Promise 的 polyfill 实现。

19.6 提取公共代码

Webpack4 中可以将多个公共模块打包一份，减少代码冗余，Webpack4 之前的版本是使用 Webpack 内置的 CommonsChunkPlugin 实现的，Webpack4 直接配置 optimization 即可。

```
module.exports = {
  optimization:{
    splitChunks:{
      cacheGroups:{
        common:{ // 应用代码中公共模块
          chunks: 'all',
          // 最小公共模块引用次数
          minChunks: 2
        },
        vendor:{ // node_modules 中第三方模块
          test: /node_modules/,
          chunks: 'all',
          minChunks: 1
        }
      }
```

```
    }
  }
};
```

第三方库代码的变更一般比较少（通过 package.json 的版本可以指定依赖版本），因此构建出来的 vendor.js 基本不用变更就可以利用浏览器的缓存机制进行缓存。

而应用代码的变更是比较频繁的，因此单独打包为 common.js，浏览器可以单独缓存，如果应用代码发生变更，浏览器只用重新下载 common.js 文件，而不用重新下载 vendor.js。

19.7 热更新

HMR（Hot Module Replacement，热更新）是 Webpack 提供的常用功能之一，它允许在运行时对模块进行修改，而无须刷新整个页面（LiveReload 需要刷新页面才能加载），这样做有以下优势：

- 保留应用状态，比如使用 Vue/React 时如果使用 LiveReload，组件状态就会全部丢失，而使用 HMR 则不会。
- 只更新变更的内容，节省开发时间。

使用以下配置即可打开内置的 HMR 功能：

```
const webpack = require('webpack');

module.exports = {
  devServer: {
    hot: true,  // 启用热加载
    contentBase: './dist',
  },
  plugins:[
    new webpack.NamedModulesPlugin(),           // 打印更新的模块路径
    new webpack.HotModuleReplacementPlugin()    // 热更新插件
  ]
};
```

19.8 本章小结

本章我们学习了 Webpack4 最常用的性能优化技术，这些优化方法对业务代码的侵入性非常小（只有按需加载优化会要求使用 import()函数进行加载）。在实际开发中，可以结合这些技术进行针对性的优化，比如开发时编译慢，可能就需要使用 HappyPack 插件进行多进程编译，以加快编译速度。

第 20 章

编写自定义 Loader

本章内容

- 自定义 Loader 的编写

在前面的章节中，我们学习了 Webpack 的基本知识、常用脚手架和性能优化等内容，虽然说大部分的开发场景社区已经提供了成熟的模块给我们使用，但是遇到特殊情况还是需要自己有独立开发自定义模块的能力，因此本章我们一起来学习如何编写自定义 Loader。

20.1 基本 Loader

Webpack 中 Loader 是一个 CommonJS 风格的函数，接收输入的源码，通过同步或异步的方式替换源码后进行输出。

```
module.exports = function(source, sourceMap, meta) {

}
```

- source: 是输入的内容。
- sourceMap: 是可选的。
- meta: 是模块的元数据，也是可选的。

需要注意的是，该导出函数必须使用 function，不能使用箭头函数，因为 Loader 编写过程中会经常使用到 this 访问选项和其他方法。

我们先编写一个基本的 Loader，完成的工作很简单，那就是把输出的字符串进行替换。

（1）新建 loader-example 目录，执行 npm 命令初始化，并安装 Webpack：

```
mkdir loader-example
cd loader-example
npm init -y
npm install webpack webpack-cli
```

（2）构建项目目录：

```
|----loader          # loader 目录
    |----replace-loader.js    # 替换字符串的 Loader
|----src             # 应用源码
    |----index.js             # 首页
|----package.json
|----webpack.config.js
```

（3）编写 loader/replace-loader.js：

```
module.exports = function(source) {
  return source.replace(/World/g, 'Loader');
};
```

本例中我们 Loader 只是简单地将源码中的"World"替换成了"Loader"。

（4）编写 src/index.js：

```
console.log('Hello World');
```

（5）编写 webpack.config.js：

```
const path = require('path');
module.exports = {
  entry: './src/index',
  target: 'node', // 编译为 Node.js 环境下的 JS，之后直接使用 Node.js 执行编译完成的文件
  output:{
    path: path.resolve(__dirname, 'build'),
    filename: '[name].js'
  },
  module:{
    rules:[
      {
        test:/\.js$/,
        use: 'replace-loader'
      }
    ]
  },
  resolveLoader: {
    modules: ['./node_modules', './loader']  // 配置 loader 的查找目录
  }
};
```

(6)编写 package.json：

```
{
  "scripts":{
    "build":"webpack"
  }
}
```

(7)执行构建：

```
npm run build
```

(8)构建完成后，执行 build/main.js：

```
node build/main.js
```

此时终端输出如下，说明我们编写的 Loader 工作正常。

```
Hello Loader
```

20.2 Loader 选项

我们使用第三方 Loader 时经常可以看到传递选项的情况：

```
{
  test:/\.js$/,
  use:[
    {
      loader:'babel-loader',
      options:{
        plugins:['@babel/transform-runtime'],
        presets:['@babel/env']
      }
    }
  ]
}
```

在 Loader 编写时，Webpack 中官方推荐通过 loader-utils 来读取配置选项，我们需要先安装它：

```
npm install loader-utils
```

给刚才编写的 replace-loader 传递一个选项，允许自定义替换的结果：

```
const loaderUtils = require('loader-utils');

module.exports = function(source) {
  const options = loaderUtils.getOptions(this);
  return source.replace(/World/g, options.text);
};
```

接下来编辑 webpack.config.js，给 replace-loader 传递选项：

```
module.exports = {
  module:{
    rules:[
      {
        test:/\.js$/,
        use:[
          {
            loader:'replace-loader',
            options:{
              text: 'Webpack4'
            }
          }
        ]
      }
    ]
  },
  resolveLoader:{
    modules: ['./node_modules', './loader']
  }
};
```

执行构建之后用 Node.js 执行 build/main.js，可以看到输出的内容已经发生了变化：

```
Hello Webpack4
```

20.3 异步 Loader

在 Loader 中，如果存在异步调用，那么就无法直接通过 return 返回构建后的结果，此时需要使用 Webpack 提供的回调函数将数据进行回调。

Webpack4 给 Loader 提供了 this.async()函数，调用之后返回一个回调函数 callback，callback 函数的签名如下：

```
function callback(
  err: Error|null,
  content: string|Buffer,
  sourceMap?:SourceMap,
  meta?: any
)
```

例如我们需要在 loader 中调用 setTimeout 进行等待，相应的代码如下：

```
module.exports = function(source) {
  const callback = this.async();
  setTimeout(() => {
    const output = source.replace(/World/g, 'Webpack4');
    callback(null, output);
```

```
}, 1000);
}
```

执行构建，Webpack 会等待一秒，然后再输出构建内容，通过 Node.js 执行构建后的文件，输出如下：

```
Hello Webpack4
```

20.4 "Raw" Loader

默认情况下，资源文件会被转化为 UTF-8 字符串，然后传给 Loader。通过设置 raw，Loader 可以接收原始的 Buffer，用于处理非文本文件（如图片）的情况。

```
module.exports = function(source) {
  assert(source instanceof Buffer);
  return someSyncOperation(source);
};
module.exports.raw = true; // 设置当前 Loader 为 raw loader, webpack 会将原始的 Buffer
    对象传入
```

20.5 读取 Loader 配置文件

babel-loader 在使用时可以加载 .babelrc 配置文件来配置 plugins 和 presets，减少了 webpack.config.js 的代码量，便于日后的维护。接下来我们编写一个 i18n-loader，通过读取语言配置文件完成语言转换。

20.5.1 项目结构

项目的目录结构如下：

```
|----loader
    |----i18n-loader.js  # loader
|----i18n
    |----zh.json         # 中文语言包
|----src
    |----index.js        # 入口文件
|----webpack.config.js
```

（1）i18n/zh.json

```
{
  "hello": "你好",
  "today": "今天"
}
```

（2）loader/i18n-loader.js

```js
const loaderUtils = require('loader-utils');
const path = require('path: ');

module.exports = function (source) {
    const options = loaderUtils.getOptions(this);
    const locale = options ? options.locale : null;

    // 读取语言配置文件
    let json = null;
    if (locale) {
        const filename = path.resolve(__dirname, '..', 'i18n', `${locale}.json`);
        json = require(filename);
    }

    // 读取语言标记 {{}}
    const matches = source.match(/\{\{\w+\}\}/g);
    for (const match of matches) {
        const name = match.match(/\{\{(\w+)\}\}/)[1].toLowerCase();
        if (json !== null && json[name] !== undefined) {
            source = source.replace(match, json[name]);
        } else {
            source = source.replace(match, name);
        }
    }
    return source;
}
```

（3）src/index.js

```js
console.log('{{Hello}}, {{Today}} is a good day.');
webpack.config.js
const path = require('path');

module.exports = {
    entry: './src/index',
    output: {
        path: path.resolve(__dirname, 'build'),
        filename: '[name].js'
    },
    target: 'node',
    module: {
        rules: [
            {
                test: /\.js$/,
                use: [
                    {
                        loader: 'i18n-loader',
                        options: { // 传递选项
                            locale: 'zh'
```

```
                }
            }
        ]
      }
    ]
  },
  resolveLoader: {
    modules: ['./node_modules', './loader']
  }
};
```

(4) package.json

```
{
  "scripts":{
    "build":"webpack"
  }
}
```

20.5.2 执行构建

构建命令如下：

```
npm run build
```

构建完毕后使用 Node.js 执行 build/main.js 输出如下：

```
你好，今天 is a good day.
```

可以看到 i18n-loader 成功读取了配置文件。

20.6 本章小结

本章简要介绍了 Webpack 中如何编写一个自定义的 Loader，权当抛砖引玉，更多的用法等待读者在实际工作中去挖掘，要想掌握 Webpack 的高级知识，Loader 是必不可少的技能，有时候如果在社区找不到合适的 Loader，就可以根据需要自己进行开发。

第 21 章

编写自定义插件

本章内容

- 自定义插件的编写

Webpack 通过 Loader 完成模块的转换工作,让"一切皆模块"成为可能。Plugin 机制则让其更加灵活,可以在 Webpack 生命周期中调用钩子完成各种任务,包括修改输出资源、输出目录等等。

本章我们一起来学习如何编写 Webpack 插件。

21.1 基本构建流程

在编写插件之前,还需要了解一下 Webpack 的构建流程,以便在合适的时机插入合适的插件逻辑。Webpack 的基本构建流程如下:

(1)校验配置文件。
(2)生成 Compiler 对象。
(3)初始化默认插件。
(4)run 阶段:如果运行在 watch 模式则执行 watch 方法,否则执行 run 方法。
(5)compilation 阶段:创建 Compilation 对象回调 compilation 相关钩子。
(6)emit 阶段:文件内容准备完成,准备生成文件,这是最后一次修改最终文件的机会。
(7)afterEmit 阶段:文件已经写入磁盘完成。
(8)done 阶段:完成编译。

21.2 插件示例

一个典型的 Webpack 插件代码如下：

```
// 插件代码
class MyWebpackPlugin {
  constructor(options) {
  }

  apply(compiler) {
    // 插入钩子函数
    compiler.hooks.emit.tap('MyWebpackPlugin', (compilation) => {});
  }
}

module.exports = MyWebpackPlugin;
```

接下来需要在 webpack.config.js 中导入这个插件：

```
module.exports = {
  plugins:[
    // 传入插件实例
    new MyWebpackPlugin({
      param:'paramValue'
    }),
  ]
};
```

Webpack 在启动时会实例化插件对象，在初始化 compiler 对象之后会调用插件实例的 apply 方法，传入 compiler 对象，插件实例在 apply 方法中会注册感兴趣的钩子，Webpack 在执行过程中会根据构建阶段回调相应的钩子。

21.3 Compiler 与 Compilation 对象

在编写 Webpack 插件过程中，最常用也是最主要的两个对象就是 Webpack 提供的 Compiler 和 Compilation，Plugin 通过访问 Compiler 和 Compilation 对象来完成工作。

- Compiler 对象包含了当前运行 Webpack 的配置，包括 entry、output、loaders 等配置，这个对象在启动 Webpack 时被实例化，而且是全局唯一的。Plugin 可以通过该对象获取到 Webpack 的配置信息进行处理。
- Compilation 对象可以理解编译对象，包含了模块、依赖、文件等信息。在开发模式下运行 Webpack 时，每修改一次文件都会产生一个新的 Compilation 对象，Plugin 可以访问到本次编译过程中的模块、依赖、文件内容等信息。

常见钩子

Webpack 会根据执行流程来回调对应的钩子，如表 21-1 所示，我们来看看都有哪些常见钩子，这些钩子支持的 tap 操作是什么。

表 21-1 Webpack 的常见钩子

钩子	说明	参数	类型
afterPlugins	启动一次新的编译	compiler	同步
compile	创建 compilation 对象之前	compilationParams	同步
compilation	compilation 对象创建完成	compilation	同步
emit	资源生成完成，输出之前	compilation	异步
afterEmit	资源输出到目录完成	compilation	异步
done	完成编译	stats	同步

21.4　Tapable

Tapable 是 Webpack 的一个核心工具，Webpack 中许多对象扩展自 Tapable 类。Tapable 类暴露了 tap、tapAsync 和 tapPromise 方法，可以根据钩子的同步/异步方式来选择一个函数注入逻辑。

- tap: 同步钩子。
- tapAsync: 异步钩子，通过 callback 回调告诉 Webpack 异步执行完毕。
- tapPromise: 异步钩子，返回一个 Promise 告诉 Webpack 异步执行完毕。

1. tap

tap 是一个同步钩子，同步钩子在使用时不可以包含异步调用，因为函数返回时异步逻辑有可能未执行完毕而导致问题。

下面是一个在 compile 阶段插入同步钩子的示例：

```
compiler.hooks.compile.tap('MyWebpackPlugin', params => {
  console.log('我是同步钩子')
});
```

2. tapAsync

tapAsync 是一个异步钩子，我们可以通过 callback 告知 Webpack 异步逻辑执行完毕。

下面是一个在 emit 阶段的示例，在 1 秒后打印文件列表：

```
compiler.hooks.emit.tapAsync('MyWebpackPlugin', (compilation, callback) => {
```

```
  setTimeout(()=>{
    console.log('文件列表', Object.keys(compilation.assets).join(','));
    callback();
  }, 1000);
});
```

3. tapPromise

tapPromise 也是异步钩子，和 tapAsync 的区别在于 tapPromise 是通过返回 Promise 来告知 Webpack 异步逻辑执行完毕的。

下面是一个将生成结果上传到 CDN 的示例：

```
compiler.hooks.afterEmit.tapPromise('MyWebpackPlugin', (compilation) => {
  return new Promise((resolve, reject) => {
    const filelist = Object.keys(compilation.assets);
    uploadToCDN(filelist, (err) => {
      if(err) {
        reject(err);
        return;
      }
      resolve();
    });
  });
});
```

apply 方法中插入钩子的一般形式如下：

```
compiler.hooks.阶段.tap函数('插件名称', (阶段回调参数) => {

});
```

21.5 常用操作

21.5.1 读取输出资源、模块及依赖

在 emit 阶段，我们可以读取最终需要输出的资源、chunk、模块和对应的依赖，如果有需要还可以更改输出资源。

```
apply(compiler) {
  compiler.hooks.emit.tapAsync('MyWebpackPlugin', (compilation, callback) => {
    // compilation.chunks 存放了代码块列表
    compilation.chunks.forEach(chunk => {
      // chunk 包含多个模块，通过 chunk.modulesIterable 可以遍历模块列表
      for(const module of chunk.modulesIterable) {
        // module 包含多个依赖，通过 module.dependencies 进行遍历
        module.dependencies.forEach(dependency => {
          console.log(dependency);
        });
```

```
      }
    });
    callback();
  });
}
```

21.5.2 修改输出资源

通过操作 compilation.assets 对象，我们可以添加、删除和更改最终输出的资源。

```
apply(compiler) {
  compiler.hooks.emit.tapAsync('MyWebpackPlugin', (compilation) => {
    // 修改或添加资源
    compilation.assets['main.js'] = {
      source() {
        return 'modified content';
      },
      size() {
        return this.source().length;
      }
    };
    // 删除资源
    delete compilation.assets['main.js'];
  });
}
```

assets 对象需要定义 source 和 size 方法，source 方法返回资源的内容，支持字符串和 Node.js 的 Buffer，size 返回文件的大小字节数。

21.6 插件编写实例

接下来我们开始编写自定义插件，所有插件使用的示例项目如下（需要安装 Webpack 和 webpack-cli）：

```
|----src
    |----main.js
|----plugins
    |----my-webpack-plugin.js
|----package.json
|----webpack.config.js
```

相关文件的内容如下：

```
// src/main.js
console.log('Hello World');
// package.json
{
```

```
"scripts":{
  "build":"webpack"
 }
}
const path = require('path');
const MyWebpackPlugin = require('my-webpack-plugin');

// webpack.config.js
module.exports = {
  entry:'./src/main',
  output:{
    path: path.resolve(__dirname, 'build'),
    filename:'[name].js',
  },
  plugins:[
    new MyWebpackPlugin()
  ]
};
```

21.6.1 生成清单文件

通过在 emit 阶段操作 compilation.assets 生成清单文件。

```
class MyWebpackPlugin {
  apply(compiler) {
    compiler.hooks.emit.tapAsync('MyWebpackPlugin', (compilation, callback) => {
      const manifest = {};
      for (const name of Object.keys(compilation.assets)) {
        manifest[name] = compilation.assets[name].size();
        // 将生成文件的文件名和大小写入 manifest 对象
      }
      compilation.assets['manifest.json'] = {
        source() {
          return JSON.stringify(manifest);
        },
        size() {
          return this.source().length;
        }
      };
      callback();
    });
  }
}
module.exports = MyWebpackPlugin;
```

构建完成后会在 build 目录添加 manifest.json，内容如下：

```
{"main.js":956}
```

21.6.2 构建结果上传到 CDN

在实际开发中，资源文件构建完成后一般会同步到 CDN，最终前端界面使用的是 CDN 服务器上的静态资源。

下面我们编写一个 Webpack 插件，文件构建完成后上传 CDN。本插件示例采用的是七牛 CDN，各位读者可以根据需要选择适合自己的 CDN 服务商。

我们的插件依赖 qiniu，因此需要额外安装 qiniu 模块：

```
npm install qiniu --save-dev
```

七牛的 Node.js SDK 文档地址如下：

```
https://developer.qiniu.com/kodo/sdk/1289/nodejs
```

开始编写插件代码：

```
const qiniu = require('qiniu');
const path = require('path');

class MyWebpackPlugin {
    // 七牛 SDK mac 对象
    mac = null;

    constructor(options) {
        // 读取传入选项
        this.options = options || {};
        // 检查选项中的参数
        this.checkQiniuConfig();
        // 初始化七牛 mac 对象
        this.mac = new qiniu.auth.digest.Mac(
            this.options.qiniu.accessKey,
            this.options.qiniu.secretKey
        );
    }
    checkQiniuConfig() {
        // 配置未传 qiniu，读取环境变量中的配置
        if (!this.options.qiniu) {
            this.options.qiniu = {
                accessKey: process.env.QINIU_ACCESS_KEY,
                secretKey: process.env.QINIU_SECRET_KEY,
                bucket: process.env.QINIU_BUCKET,
                keyPrefix: process.env.QINIU_KEY_PREFIX || ''
            };
        }
        const qiniu = this.options.qiniu;
        if (!qiniu.accessKey || !qiniu.secretKey || !qiniu.bucket) {
            throw new Error('invalid qiniu config');
        }
```

```js
    }
    apply(compiler) {
        compiler.hooks.afterEmit.tapPromise('MyWebpackPlugin', (compilation) => {
            return new Promise((resolve, reject) => {
                // 总上传数量
                const uploadCount = Object.keys(compilation.assets).length;
                // 已上传数量
                let currentUploadedCount = 0;
                // 七牛 SDK 相关参数
                const putPolicy = new qiniu.rs.PutPolicy({ scope: this.options.qiniu.bucket });
                const uploadToken = putPolicy.uploadToken(this.mac);
                const config = new qiniu.conf.Config();
                config.zone = qiniu.zone.Zone_z1;
                const formUploader = new qiniu.form_up.FormUploader()
                const putExtra = new qiniu.form_up.PutExtra();
                // 因为是批量上传，需要在最后将错误对象回调
                let globalError = null;

                // 遍历编译资源文件
                for (const filename of Object.keys(compilation.assets)) {
                    // 开始上传
                    formUploader.putFile(
                        uploadToken,
                        this.options.qiniu.keyPrefix + filename,
                        path.resolve(compilation.outputOptions.path, filename),
                        putExtra,
                        (err) => {
                            console.log(`uploade ${filename} result: ${err ? `Error:${err.message}` : 'Success'}`)
                            currentUploadedCount++;
                            if (err) {
                                globalError = err;
                            }
                            if (currentUploadedCount === uploadCount) {
                                globalError ? reject(globalError) : resolve();
                            }
                        });
                }
            })
        });
    }
}

module.exports = MyWebpackPlugin;
```

在 Webpack 中需要给该插件传递相关配置：

```js
module.exports = {
    entry: './src/index',
```

```
    target: 'node',
    output: {
        path: path.resolve(__dirname, 'build'),
        filename: '[name].js',
        publicPath: 'CDN 域名'
    },
    plugins: [
        new CleanWebpackPlugin(),
        new QiniuWebpackPlugin({
            qiniu: {
                accessKey: '七牛 AccessKey',
                secretKey: '七牛 SecretKey',
                bucket: 'static',
                keyPrefix: 'webpack-inaction/demo1/'
            }
        })
    ]
};
```

编译完成后资源会自动上传到 CDN，这样前端只用交付 index.html 即可。

21.7 本章小结

至此，Webpack 相关常用知识和进阶知识都介绍完毕，需要各位读者在工作中去多加探索，Webpack 配合 Node.js 生态，一定会涌现出更多优秀的新语言和新工具！